COMPUTER
COMMUNICATIONS

ASPECTS OF
INFORMATION TECHNOLOGY

This series is aimed primarily at final year undergraduate and postgraduate students of Electronics and Computer Science, and provides an introduction to research topics in Information Technology which are currently being translated into teaching course material. The series aims to build bridges between foundation material covered in the first two years of undergraduate courses and the major research topics now attracting interest within the field of IT.

The format of the series is deliberately different from that of typical research reference works within the fields of interest. Depth of coverage is restricted in favour of providing a readable and comprehensible introduction to each topic, and to keep costs within the requirements for a course textbook. Nevertheless, each book provides a comprehensive overview and introduction to its subject, aimed at conveying the key elements in an attractive and clear fashion.

Each chapter is terminated by a summary itemizing the key points contained within, and problems and exercises are provided where appropriate to enable the student to test his knowledge. For the serious student, each book contains a comprehensive further reading list of texts and key reference papers covering the field, as well as giving an indication of those journals which publish within the area.

Series editors: **A C Downton**   *University of Essex*
                      R D Dowsing   *University of East Anglia*

COMPUTER COMMUNICATIONS

K. G. Beauchamp
Former Director of Computer Services
University of Lancaster

Van Nostrand Reinhold (UK) Co. Ltd

'The cardinal fact in human history during
the last thirty centuries has been the scope,
pace and precision of intercommunication.
Everything else is subordinated to that.'

H. G. Wells, 1935

First published in 1987 by
Van Nostrand Reinhold (UK) Co. Ltd.
Molly Millars Lane, Wokingham,
Berkshire, England

Typeset in 10/12pt Times by
Witwell Ltd, Liverpool

Printed and bound in Great Britain by T.J. Press
(Padstow) Ltd, Padstow, Cornwall

Beauchamp, K.G.
    Computer communications.—(Aspects of
    information technology).
    1. Computer networks
    I. Title   II. Series
    004.6   TK5105.5

    ISBN 0–278–00012–6

CONTENTS

PREFACE

The convergence of electronic communications and the digital computer within the last decade has resulted in the growth of a new communications industry supported by a well-defined body of theoretical and practical knowledge concerning the new techniques. An area of especial interest is the subset of ideas and techniques available for linking computers together, forming a network used to share resources or tasks. The subject of computer networking is a rapidly changing one, particularly having regard to the problems involved in linking together small computers over a limited geographical area – the local area network. Industry is moving towards standardization in this, as well as in other areas of networking, either though formal international agreement or by setting their own *de facto* standards. The range of applications for computer networks is, however, a wide one and there is room for a number of different techniques to be available so that a choice can be made for a given computer communication requirement. It is the intention with this book to provide a profile of the present range of these activities with sufficient basic information to enable newer developments to be understood as they come along.

This book is intended to provide material suitable for a one-year introductory academic course in the subject. To follow this the student requires very little in the way of previous knowledge. A minimal understanding of how computers work is assumed but no previous exposure to electronic communication will be needed. A number of sample questions are included at the end of each chapter. Numeric answers are given at the end of the book and some notes to indicate ways in which the non-numeric questions could be attempted.

Within the limited space available, an exhaustive treatment of the application of these communication topics is not possible. The final chapter provides a resumé of current literature on the subject and reference to the key articles mentioned in earlier chapters. It is hoped that these will provide a sufficient guide to the wider and detailed questions arising from the text.

I would like to acknowledge the assistance of many people and organizations during the preparation of this book. In particular I would like to thank Dr R.D. Dowsing of the University of East Anglia and Dr. A.C. Downton of the University of Essex for reading the text in draft form and for providing much useful comment and suggestions on the material. I also wish to thank Professor C.K. Yuen and his staff at the University of Singapore (where much of the writing was carried out), Professor M. Wells of the University of Leeds for information on the JANET project, the library staff at the Institution of Electrical Engineers in London and a number of organizations who have provided much detailed information on their products. These including Dr. J. Howlett of International Computers Ltd, Mr R.C. Hooper of British Telecom and Mr V. Teacher of Standard Telephones and Cables. Finally, I wish to record my thanks to Mah Chui Yoke Inger for her work in the preparation of the various drafts of the text.

K.G. Beauchamp
Singapore and
Lancaster 1986

ABBREVIATIONS

A T & T   American Telegraph & Telephone Company
ABM   asynchronous balanced mode
ACK   acknowledgement
AM   amplitude modulation
ARM   asynchronous response mode
ARPA   (Defence) Advanced Research Projects Agency (USA)
ASCII   American standard code for information exchange
ASK   amplitude-shift keying
B   bandwidth
BIOS   basic input/output system
bps   bits per second
CASE   common applications service elements
CBX   computerized branch exchange
CCITT   International Consultative Committee for Telephones and
            Telegraphs
CCS   common-channel signalling
CFR   Cambridge fast ring
CRC   cyclic redundancy check
CSMA/CA   carrier sense multiple access with collision avoidance
CSMA/CD   carrier sense multiple access with collision detection
dB   decibels
d.c.   direct current
DCE   data circuit terminating equipment
DLT   digital time termination unit
DPSK   differential phase-shift keying
DSB   double side-band
DTE   data terminating equipment
EARN   European Academic Research Network
EBCDIC   Extended Binary Coded Decimal Interchange Code

FCS	frame check sequence
FDM	frequency division multiplexing
FM	frequency modulation
FSK	frequency-shift keying
FTAM	file transfer applications and management
FTP	file transfer protocol
HDB3	high-density bi-polar code (No 3)
HDLC	High Level Data Link Control
HM	hybrid modulation
HSLN	High-speed local network
Hz	Hertz (cycles per second)
IA5	International coding alphabet
ICP	Interconnection protocol
IDN	Integrated digital network
IEE	Institution of Electrical Engineers
IEEE	Institution of Electrical and Electronics Engineers
ILD	injector laser diode
IMP	interface message processor
IPMS	interpersonal message system
IPSS	International Packet-switched service
ISDN	Integrated Services Digital Network
ISO	International standard organisation
j	square root of –1
JANET	joint academic network
JNT	Joint Network Team
JPSE	JANET packet-switching exchange
JTMP	job transfer and manipulation protocol
LAN	local area network
LAP-B	link access protocol — balanced
LED	light-emitting diode
LLC	logical link control
lsb	lower side-band
MAC	medium access control
MAP	manufacturing applications protocol
MAU	multistation access unit
MHS	message handling system
MMFS	manufacturing message service standard
MPX	multiplexer
MTA	message transfer agent
NAK	negative acknowledgement
NETBIOS	network BIOS
NCC	Network Control Centre
NIFTP	network-independant file transfer protocol
NITS	network-independent transport service
NMU	Network Management Unit

NOC	Network Operations Centre
NRM	normal response mode
NRZ	non-return to zero
NT1	network terminal (No 1)
OSI	open systems interconnection
P/F	poll/final (bit)
PABX	private automatic branch exchange
PAD	packet assembler/disassembler
PAM	pulse amplitude modulation
PC	personal computer
PCM	pulse code modulation
PDN	public data network
PM	pulse modulation
PSDN	public switched data network
PSK	phase-shift keying
PSS	Packet-switched Service
PSTN	Public-switched telephone network
QAM	quadrature amplitude modulation
QPSK	quadrature phase-shift keying
RJE	remote job entry
RTS	reliable transfer service
RZ	return to zero
SAP	service access point
SDLC	Synchronous Data Link Control
SNDCF	subnetwork-dependent convergence facility
SNR	signal-to-noise ratio
SNICF	subnetwork-independent convergence facility
SSB	single side-band
SSB-SC	single side-band suppressed carrier
TDM	time division multiplexing
TIP	terminal interface processor
TOP	technical and office protocol
TSE	terminal-switched exchange
UA	user access
usb	upper side-band
VAN	value added network
VSB	vestigial side band
WAN	wide area network

1

PRELIMINARIES

1.1 HISTORICAL DEVELOPMENT

The origins of electrical communications may be found from the works of Wheatstone and Cooke in the early 1800s. The first electric telegraph, demonstrated by these inventors to the directors of the London to Birmingham Railway in 1837, relied on the interaction of an electric current with a pivoted magnetic needle, which moved to indicate the presence and direction of a current in a transmission line connecting the sending apparatus and the receiving apparatus. By using a number of transmission lines, each connected to a separate needle, a code could be worked out which related the deflections of the needle to letters and numbers constituting a transmission code. This system required some skill in operation and was confined to the burgeoning railway system which urgently needed a reliable and rapid signalling method to ensure the safety of the travelling public.

The early telegraph engineers quickly discovered restrictions in operation of this equipment, principally in signal attenuation with distance and induced electrical interference, both of which imposed limitations on widespread application outside the railway system.

A much improved system was advocated by Samuel Morse in 1835 which, through an ingenious combination of an electric magnet and an iron plunger carrying a writing device, was able to mark a moving strip of paper with long and short coding marks corresponding to the duration of the current sent along a transmission line. The code he used consisted of two binary variables — the presence or absence of a current and two alternative duration periods for this current. The **Morse code** he evolved was enormously successful and enabled a more general and easily applicable transmission scheme to be put into effect for message transmission. Towards the end of the century Marconi, following the work of Hertz and others, also applied the coding methods of Morse to radio transmission using a spark transmitter.

It is interesting to compare the operational speeds of these early binary systems with present-day digital transmission. Assuming a 3:1 ratio between a mark and a space and considering the longest code sequence to be five spaces, then equating a digital bit with a Morse code space, a complex character would occupy 15 bits. A good Morse operator would be capable of a working speed of about 30 words per minute; a word being considered to have an average of five characters. This gives a speed of 37 bits per second (bps) which compares quite favourably with the lowest commercial speed available for public transmission, namely the Datel services operating at 200 bps and the Telex service at 80 bps.

Although valuable for simple message transmission, business and public usage demanded more than a simple and flexible coding system in a communication device. Rapid dialogue between users was expected and a system which did not need message translation and the intervention of a skilled operator to work the equipment was necessary.

1.1.1 The telephone network

Alexander Graham Bell was to provide the basic technique to match these requirements, with his invention of the telephone in 1876. From this has evolved a complex telecommunications network which, in its present form may be described as a distributed intelligent machine, arguably the biggest machine ever made by man. It presently links some 600 million users worldwide, each of whom are capable of communicating with any other, irrespective of geographical location.

Initially, however, the problem of connecting even a small group of N users proved a daunting task, since each addition to the number of connected users meant a further N^2 lines would need to be provided. The solution was to re-introduce the operator, this time to connect the calls through a local switchboard, and thus to establish a **switched network**. The function of such a network is to involve switching centres, or **nodes** to carry out the processes of control and routing and to establish transmission links for connecting external machines to the network. The basic switching functions, which are the same today for telephone or data traffic (although the terms used may differ), are shown in Table 1.1.

The early manual exchanges were very simple with the control vested in a trained operator. As the number of lines proliferated, the number of switchboards increased, imposing a practical limit on the amount of interaction and volume of telephone traffic that could be realized. Indeed, with the volume of today's telephone traffic Young[1] has speculated that a sizable fraction of the Earth's population would be needed by now to operate all the manual exchanges required!

The first fully automatic exchange was invented by Almon Strowger towards the end of the nineteenth century and installed in La Portee, Indiana

Table 1.1

Basic switching functions

Function	Action
1	Detection of calling condition (off-hook)
2	Initial path set up
3	Signalling from the sender
4	Call set up
5	Alerting (ringing)
6	Answer detected from receiver
7	Conversation phase
8	Clearance detected (on-hook)
9	Call clear down

in 1892[2]. In his system, the first of many others essentially like it, the switches were operated directly and step-by-step by pulses of current were sent along a transmission line. The line was thus used to carry not only the speech signal information but also signalling information, a system now referred to as **in-band signalling**. To separate these two forms of signalling along the same wire, a carefully worked out set of rules or signalling **protocols** needed to be evolved to carry out the basic switching functions shown in Table 1.1, with the line reserved for speech transmission only during phase 7.

1.1.2 Equalization

Following the successful introduction of the Strowger system, the increase in telephone usage led to new methods of multi-channel communication which made better use of the available bandwidth. Early experiments with open line telegraph transmission had shown that the signal attenuation (see Appendix to this chapter) with distance from the transmitting location is not constant but a function of frequency, so that the higher frequency components of the transmitted signal are reduced in signal amplitude at a faster rate than the lower frequency components. This is shown in Fig.1.1a. To overcome this, the characteristics of the signal applied at the sending end are modified such that the higher frequencies have greater amplitude than the lower frequency components. This process is known as **equalization** and results in the more level frequency response shown in Fig.1.1b, thus permitting the length over which the signal may be transmitted without distortion to be considerably increased.

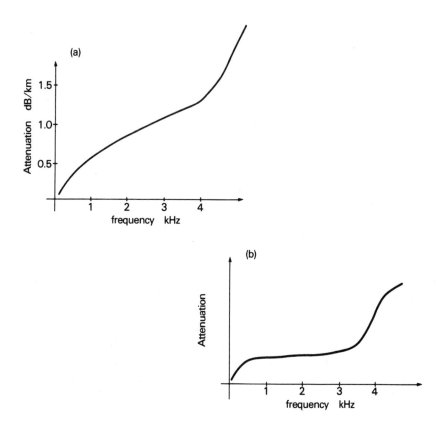

Fig. 1.1 (a) Frequency distortion with open line transmission.
(b) With Equalization applied.

1.1.3 Multi-channel communication

With equalization applied, the effective bandwidth for a simple open wire transmission system was found to be more than adequate to support a single telephone channel over considerable distances. There were, however, on a given route, a great number of communication channels to support, and long-distance communication for the **trunk routes** established between exchanges demanded many thousands of pairs of wires if this concept of **space division switching** was to be applied, i.e. each pair of wires arranged to carry one speech circuit only.

To make maximum use of each cable it is desirable to be able to transmit more than one conversation over a single pair of wires, and so methods were devised to permit their separation at the distant end. One way of carrying this out is through a process of **frequency division multiplexed** (FDM) switching.

Each speech transmission is superimposed on a separate **carrier signal** through a process of **modulation**. By choosing a different carrier frequency to convey individual speech signals, each conversation is effectively shifted to a different part of the frequency spectrum.

Amplitude modulation was first demonstrated in an early radio transmission by Dr Reginald Fassenden in the USA through a historic broadcast on Christmas Eve 1906. The technique was quickly appreciated by telegraph engineers and an adaptation of this principle served to solve the problem of carrying many speech signals over a single line in telephony transmission.

Amplitude modulation is a process of varying the amplitude of the sinusoidal carrier wave by the amplitude of the modulating signal, Fig.1.2 The unmodulated carrier wave (b) has a higher frequency than the modulating signal (a) and its normally constant peak value is varied in

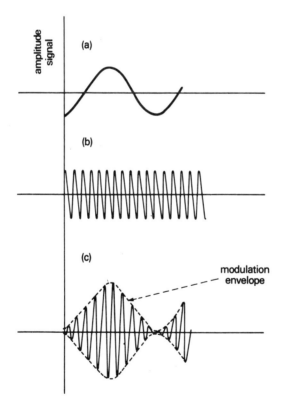

Fig. 1.2 Amplitude modulation. (a) Modulation waveform; (b) unmodulated carrier waveform; (c) modulated carrier waveform.

accordance with the instantaneous value of the modulated carrier signal to produce the modulated signal shown in (c). It can be easily demonstrated that when a sinusoidal carrier wave of frequency f_c Hz is amplitude modulated by a sinusoidal modulating signal of frequency f_m Hz, then the modulated carrier wave will contain three frequencies; the original carrier frequency f_c and two *side-band* signals, $f_c + f_m$ and $f_c - f_m$. The information content is the same for both side-bands and only one need be transmitted. Note, however, that the bandwidth occupied by the transmitted signal in the upper side-band extends from $f_c + 300$ to $f_c + 3400$ Hz instead of the 300 to 3400 Hz of the original modulating signal. This is seen in Fig. 1.3a which shows that the position in the frequency spectrum occupied by the speech signal has been moved to a new position close to f_c. By choosing a different carrier frequency for each telephone signal, then a number of such modulated **single side-band** signals can be combined and transmitted over a single communications channel without losing their separate identity, Fig.1.3b.

At the receiving end, the combined signal needs to be modulated again, in turn with a set of carrier frequencies identical to the set used for the original modulation process, so as to derive the separate modulated carrier frequencies. Recovery of the original speech waveforms from these individual carrier signals is obtained by effectively limiting the negative excursions of the

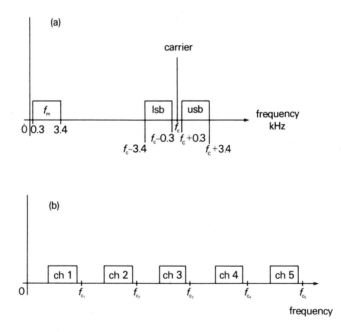

Fig. 1.3 Frequency spectrum: (a) of a single modulated signal; (b) of a group of single side-band modulated signals.

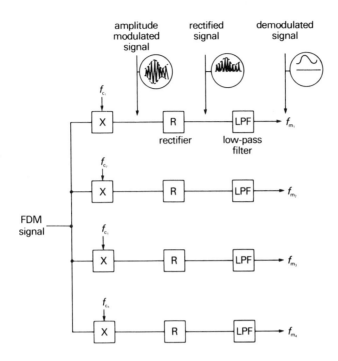

Fig. 1.4 Process of FDM signal recovery.

modulated signal, shown in Fig. 1.2c, (**rectifying** the signal) and averaging the resulting waveform. The complete process is shown in Fig. 1.4.

The first carrier telephone system in commercial service, providing four additional circuits over one pair of wires, was connected between Baltimore and Pittsburgh in 1918. This was quickly followed by others on both sides of the Atlantic, the later exchanges carrying over 600 simultaneous speech conversations on a single coaxial cable.

1.1.4 Electronic exchanges

A steady increase in the number of telephone subscribers and the heavy maintenance requirements for electro-mechanical switches led to a search for an electronic equivalent. Initially, problems associated with the high voltages used for power feeding and ringing in the existing network made it difficult for electronic devices, particularly those using semiconductor devices, to co-exist on the same network. However, by providing an interface unit at the user's line entry to the exchange, these high voltages could be deflected from the switching equipment and so permit the lower voltage transistors and later, solid state equipment, to be installed.

The first operational all-electronic exchange, replacing the slower mechanical Strowger exchanges, came into operation in 1964 in the New Jersey town of Succasunna, although the British Post Office already had an experimental system undergoing evaluation at this time [3]. The American system was called the ESS-1 (No. 1 Electronic Switching System) and had a stored memory that contained the program switching logic and a working memory holding the temporary information needed to process the call. Although not completely digital (it contained reed-relays that made the actual speech connections under electronic control), it was the first of a series of digital switches that made their appearance shortly afterwards. In Britain a similar system, the TX-E2. made its appearance two years later and a completely digital system (System X), using packet switching, came into use in 1979[4].

To match the performance of these newer exchanges, coaxial cables were laid between them which, due to their wider bandwidth capability, were able to convey very many more channels compared with multi-wire connections. More recently, fibre optic cable is being installed between all the trunk exchanges in the UK to carry the increased digital traffic for multi-chanel speech and data.

1.2 DIGITAL COMMUNICATION

A new dimension was added to electrical communication when the need for information exchange between computers and remote terminals began to materialize in the early 1960s. Until then the telephone network system was used almost entirely for speech communication and was designed specifically for this purpose. Whilst the national telephone network provided a convenient vehicle for digital data transmission there were a considerable number of problems to overcome, not least the inability of the telephone network to carry a d.c. signal level, and electrical interference which, although disregarded in speech communication, poses a severe problem for digital transmission.

The problems were met by the development of the **modem** which enables digital signals to be converted into analogue form for transmission over the telephone network. Although not ideal, the attachment of this comparatively simple and cheap device to a normal telephone line has enabled the existing telecommunications network to play a significant part in the connection of computers to remote terminals and to other computers. A severe limitation to the attachment of modems to the **public telephone service** is the effect of noise arising from relay switching, cross-talk and induced electrical interference. Tests carried out on a typical line operating at 1200 bps have shown that over 80% of transmission errors arise from a noise 'burst' spanning 2 digital bits and that the number of longer bursts of interference falls dramatically beyond a length equivalent to about 8 bits [5].

Whilst not generally implemented for modem connected systems, this type of line transmission error analysis has led to the consideration of **error correcting codes** which are effective in dealing with small error groups. They function through the addition of extra bits to the data being transmitted, followed by sophisticated forms of data analysis at the receiving end.

Due to the presence of electrical noise on the public telephone circuits, carrier systems employing modems for data transmission have generally relied on the use of private circuits (**leased lines**) to connect both ends of the transmission link. This type of operation accounts for the majority of modem connected circuits in use today.

1.2.1 Digital telephony

With the availability of digital logic gates in microelectronic form came the opportunity to apply **time division multiplexed** (TDM) switching as a method of carrying a number of speech signals over one transmission line. Here, each signal is sampled at predetermined intervals and the samples from a number of signals collected together for transmission over the transmission line. This is shown in Fig. 1.5.

TDM proved cheaper than FDM but was not put into service initially due to difficulties in integrating the new equipment with existing systems[6]. When combined with digital transmission of the TDM samples, however, in a system known as **pulse code modulation** (PCM), the method proved extremely effective. Briefly, in PCM the analogue samples shown in Fig. 1.5b are quantized into a limited number of discrete levels (eight levels are usually sufficient for speech transmission) and each level defined as a particular digital signal. The binary digits corresponding to the level are transmitted and interleaved with digits related to other sampled and digitized speech signals. The PCM/TDM transmission is separated out into its constituent pulse trains at the receiving end and conversion of the digital to the analogue (speech) form of the signal takes place. The method proved effective on economic grounds but it was realized that the use of digital techniques for speech would also provide a basis for combing non-speech services, particularly data, on the transmission network. As a result, PCM techniques were soon brought into operation for most of the inter-exchange trunk routes[7].

1.2.2 Advantages of digital transmission

Despite the problems of impulsive noise, digital transmission is now the preferred method of transmission for the telecommunications network, gradually supplanting analogue methods in all major areas of communication. It is anticipated that the British trunk network will be converted

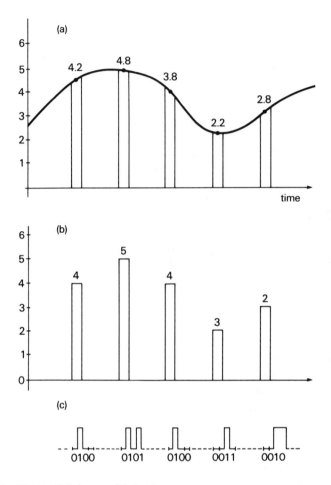

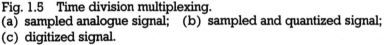

Fig. 1.5 Time division multiplexing.
(a) sampled analogue signal; (b) sampled and quantized signal;
(c) digitized signal.

completely to digital operation by 1992[8]. The most important reasons for preferring digital methods are:

a **Digital technology.** Large scale integration technology has caused a continuing drop in the cost of digital circuitry. Analogue equipment has not shown a similar drop.

b **Data integrity.** Using repeaters rather than amplifiers, the effects of noise etc. are not cumulative (see section 3.3). Hence it is possible to transmit over longer distances with digital, rather than analogue techniques.

c **Capacity utilization.** It is now economic to build transmission links of very high bandwidth. To use these effectively, it is necessary to make extensive use of multiplexing. This is easier and cheaper to do using digital technology.

d **Security and privacy**. Encryption techniques are easier to apply to digital data.

e **Integration**. By treating both analogue and digital data digitally, all signals have the same form and economies of scale can be achieved. A recent development is the progress towards an international standard for the **Integrated Services Digital Network** (ISDN) which promises to have far-reaching effects in information transmission.

1.2.3 Computer control

The use of computers for the complex control functions required in tele-communications switching, demanded extremely high standards of reliability in both hardware and software. To achieve this, use is made of stored program control, load sharing and duplication of the control facilities so that automatic changeover can take place in the event of an operational fault[9]. The availability of low cost microprocessors has also led to forms of devolution of control functions in switching exchanges and this also improves the reliability of the network.

It is also possible, with modern computer control, to separate out the signalling functions from the transmitted message or speech signal. This is known as **out-band signalling** or **common-channel signalling** and is a feature of the ISDN convergence techniques which we discuss in Chapter 9. Using out-band signalling, the number of messages that can be transmitted is increased and the signalling overheads reduced substantially.

1.2.4 Transmission of digital data

To transmit digital data, two different voltage levels are employed to represent binary 0 and 1. The smallest interval between these transitions, T, is known as the **signalling interval** or bit time and its **transmission rate**, $1/T$, referred to as so many bits per second (bps). The faster the transmission rate, the higher is the frequency bandwidth needed within the transmission media and communications equipment. For this reason, transmission rate is a major defining value in communications hardware, which is designed to offer an upper transmission rate of so many 1000 bps or kbps.

The term **baud** is sometimes applied (incorrectly) as a measure of transmission rate. However, the baud (after the nineteenth-century engineer, Baudot) is actually a measure of the **symbol transmission rate** and will depend on the nature of the digital encoding employed. For a multi-level signal having more than two levels, the bauds and bps will be different.

To determine the actual frequency bandwidth occupied by a rectangular pulse train (e.g. a sequence of binary 0 and 1 levels), we need to carry out **Fourier analysis** on the narrowest pulse in the sequence. What this would show is that a (theoretically infinite) series of harmonically related cosinusoidal components is necessary to represent the waveform in the frequency domain. A more useful way of relating frequency bandwidth requirements with the bps is to carry out a process of **harmonic synthesis** using a limited number of terms. This is illustrated graphically in Fig. 1.6.

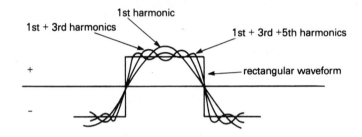

Fig. 1.6   Harmonic synthesis of a rectangular waveform.

Commencing with a term, cos $(2\pi t/T)$, as the fundamental component, a third harmonic component, $-1/3 \cos(2\pi 3t/T)$ is added. This has the effect of broadening the fundamental waveform as shown, whilst introducing a wavy flat top to the synthesized pulse. The addition of further odd harmonics from the N-term series viz:

$$x(t) = \cos(2\pi t/T) -1/3 \cos(2\pi 3t/T) + 1/5 \cos(2\pi 5t/T) - 1/7 \cos(2\pi 7t/T) +. . . + 1/N \cos(2\pi Nt/T)$$

produces a summation which approaches closer and closer to a rectangular waveform. A reasonable synthesis is to consider components up to the fifth term so that 1 bps would equal approximately 5 Hz.

Bandwidth is one of the two primary resources required for effective communication. The other is power, i.e. the strength of the signal defined in terms of the product of a current injected into the transmission media at a given voltage level. Often this is measured in terms of the product of the square of the current flowing times the impedance of the load into which it flows (see Appendix to this chapter).

In a specific communications channel one resource may be more valuable than the other. With most forms of line or computer network transmission, for example, communication bandwidth is limited and the power level is made high. In a space satellite communications system the available bandwidth can be extremely wide but the power received will be low. Equipment design will take these factors into account.

Signal power is related to the quality of transmission and is often expressed by referring to the noise characteristic of the media, the **signal-to-noise power ratio** (SNR). Increasing the signal power reduces the effect of channel noise so that the information is received more accurately or with less uncertainty. A large SNR also allows transmission over a longer distance. In general, SNR and bandwidth are exchangeable and this has important consequences in the design of data transmission systems as we will see later.

1.3 COMMUNICATIONS PROBLEMS

From the historical treatment given earlier we may observe that the growth of electrical communications, including the more recent requirements for digital computer linkage, is seen as a succession of problems encountered and overcome with consequent effect on the technology used.

What are the communications problems facing us at the present time?

Apart from the need to improve the accuracy of data transmission and to seek better ways of relaying ever larger quantities of data over a limited bandwidth medium, there are a number of requirements which can only be met by through integration of the various separate networks presently in service. For such a conglomerate service these could include:

a **Digital operation.** All signals (speech, data, video, facsimile etc.) converted to digital form.

b **Standardised protocols.** All services to conform to a standard set of protocols having international agreement and based on a multi-layer mode of operation such that the simplest activities will only make use of the lower levels in the standard.

c **Wide bandwidth.** Capability to carry wideband signals such as would be required for multi-channel speech/data communications and video signals.

d **Low error rates.** The error rate made adjustable to the type of service required.

e **Flexible connectivity.** To permit end user connection to local communications subsystems (networks), selective broadcast and conference facilities. Also authorized connection to available public and private data bases through the higher protocol systems.

f **Mailing services.** Mailbox facilities extended as a global service to include telex, viewdata, document transmission and stored speech messages.

g **Control and signalling.** Including such features as encryption, billing and access security checks.

Current technology is able to provide much of this level of convergence and its realization is expected to form the main area of telecommunications development in the next few decades.

At present there is no one technique or set of techniques which can provide all these kinds of services to the user. Even if suitable systems could be demonstrated, it will still be necessary to achieve a satisfactory level of inter-national agreement. Instead we have, in effect, a series of **overlay networks** which are separate networks providing a means of extending the national network to include a variety of services. Examples are digital packet **wide area networks** (WAN), the telex network, videotext networks, the ISDN and very many private **local area networks** (LAN). This is illustrated in Fig. 1.7. Links between networks can be provided by **gateway exchanges** as shown, which are able to adopt to different bit rates and to provide conversion between the different protocols in use.

A major purpose of this book is to describe the development of several of these overlay networks considering particularly those which are relevant to the process of communication involving the computer as the source, destination or control device for the information transmitted.

1.4 AN OUTLINE OF THE BOOK

This chapter has reviewed the basic problems in electrical communication through the historical background of continuous (analogue) communications and the benefits conferred by adopting digital transmission methods. Data transmission is principally the province of computer-to-computer communi-cation and calls for precise methods of device interaction if efficient transfer of information is to be obtained. The value of computer control in the transmission procedure and the role of protocol design is illustrated for the public telecommunications network. Techniques for partitioning the communications bandwidth to permit multi-user and multi-channel communications are introduced.

The following four chapters constitute a tutorial in the principles of digital communication methods. An exhaustive examination is not aimed for. Instead, the emphasis is on the understanding of basic principles with references to specialized texts where detailed information can be obtained.

Chapter 2 is concerned with the operation and function of continuous communications systems using a high-frequency carrier system. Methods of modulation are described for analogue systems and an introduction to their use in the conveyance of digital signals given. The operation of some of the equipment used in digital communications over telephone lines is described with particular reference to modern practice in modem design and multiplexing methods.

Chapter 3 commences with a discussion of the various types of media that

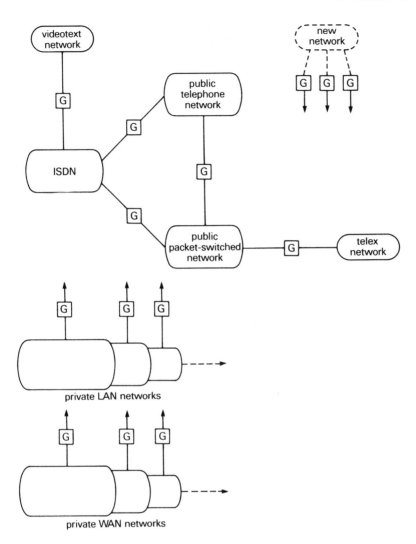

Fig. 1.7 Overlay network (G = gateway).

can be used. Emphasis is placed on current developments in fibre optics which are likely to dominate future network development. It continues with an examination of the essential conditions required for the transmission of binary information and the signal impairments that can occur.

The next two chapters, 4 and 5, consider the way in which the message or a series of messages can be formed for digital transmission and the coding methods employed. In particular, a study is made of the techniques used to ensure that the message is sent and received without error or conflict with other messages being transmitted over the same medium.

The remaining chapters consider the practical implications of computer communications techniques presented here as a merger of digital communication methods and computer control.

Chapter 6 provides a review of networking techniques considering the principles of switched networks and broadcast networks. A discussion of the advantages and disadvantages of differing network topologies for local and wide area networks is given. The practical application of packet-switched systems is considered, with particular emphasis on the public switched networks together with several value added networks which are currently in use. It concludes with an introduction to the basic problems of contention and data removal present with one or the other of these topologies.

Chapter 7 considers the local area network and discusses the various techniques in use at the present time. This is an area where recent development has been particularly rapid and where a stable technology has yet to be established. It concludes with a review of a number of different LANs currently available with brief descriptions of their operation and advantages for different forms of data traffic.

The use of computers for high-speed control of transmitted data is made possible by the application of a set of rules, known as protocols, which are the subject of Chapter 8. This is at the heart of modern methods of network technology and can be a subject of some complexity. The treatment given here is such as to enable the essentials of the methods to be understood and compared. Reference is made throughout to existing or developing standards, noting those likely to achieve international recognition in the near future.

Finally, in Chapter 9 an attempt is made to look at the most recent of the efforts made to apply digital communication techniques in the integration of a wide variety of requirements using a single communication channel. These include ISDN, MAP, TOP and electronic messaging systems. The recent development of the Integrated Services Digital Network (ISDN) as a mechanism for the convergence of all our information transmission needs over a global transmission path is perhaps the most far reaching of all current developments in information technology. If successful, this will not only coalesce much of our existing technology into a single goal, but is likely to open new fields in information transfer and to fundamentally affect the way in which society organizes itself in its day-to-day life.

PROBLEMS
Chapter 1

P1.1 What is the difference between **in-band signalling** and **out-band signalling**, applied to the public communication services?

P1.2 Define the meaning of the following terms:
a space division switching;
b frequency division multiplexed switching;
c time division multiplexed switching.

P1.3 In a given transmission system the strength of the signal applied at the sending end is at a level of 500 mV. The noise at the receiving end has an average level of 20 mV irrespective of whether there is a signal on the line or not.

If the signal attenuation along the line is 1 dB/km, calculate the maximum effective length of line that can be used before the signal-to-noise level becomes equal to 20 dB. (Assume the same electrical impedence at both ends of the line.)

P1.4 In communication systems it is often convenient to express absolute power levels in terms of decibels relative to some fixed power reference. If the reference level is 1 mW, prepare a table of absolute power levels (call these d Bm) for 1 μW, 10 μW, 100 μW, 1 mW, 1 W, and 10 W.

APPENDIX: THE DECIBEL NOTATION Chapter 1

Signal attenuation in information transmission is usually expressed as **relative attenuation** between the power levels at the input and output ends of a transmission system and is stated in **decibels** defined as:

$$\text{Attenuation} = 10 \log_{10} \frac{P_{in}}{P_{out}} \tag{1.2}$$

and where the electrical impedance of both ends of the line is the same, this may be expressed in terms of the input/output voltage levels as:

$$\text{Attenuation} = 20 \log_{10} \frac{V_{in}}{V_{out}} \tag{1.3}$$

2

CARRIER SYSTEMS

When the need arose in the early 1960s for communication at a distance with digital computers, the means were readily available. The **public switched telephone network** (PSTN) was at a high level of development and provided a convenient vehicle for data transmission. Facilities existed to allow concurrent communication over several channels by utilizing **high-frequency carrier modulation** and a process of **frequency division multiplexing**. A hierarchy of local trunk and international exchange switches were in place to convey the carrier-borne signals over considerable distances directly to the customers' premises.

The PSTN was, however, designed with the transmission of speech-only signals in mind with little or no regard paid to their variation in phase and amplitude with frequency. In addition, the speech circuits have no d.c. path (due, for example, to the presence of a.c. coupled transmission circuits) and a limited channel bandwidth of about 300 – 3400 Hz. Such a medium is unsuitable for direct use in the transmission of digital signals, whose characteristics consist of rectangular pulses producing spectral components from d.c. to several hundreds of kilohertz. Further, the method of conveying the telephone signals over long distances involves a switching process which can introduce severe transient noise in the path between the transmitter and receiver, seriously inhibiting the working of any digital equipment connected to it.

As we saw in the last chapter, these difficulties were overcome by the development of the modem which has enabled digital signals from the computer to be transmitted over the network. To improve the digital transmission characteristics, permanently connected private circuits, **leased lines**, became available which avoided route switching during a given transmission and at the same time permitted the incorporation of permanent correction for line circuit deficiencies.

The use of such private circuits became the popular method not only due to the better digital performance that could be obtained but also because most requirements for data communications were for dedicated use within a given private company requiring high data traffic density and having no need for links to other systems.

In recent years, the considerable developments in computers and their potential as an information tool has resulted in a revolutionary change in communications methods. It is now possible to deploy completely digital transmission methods involving more efficient multiplexing techniques, transmission of data in clearly defined 'packets' of information and the routing of such packets under supervisory computer control. We are also beginning to witness the convergence of speech/data/vision signals which, having been encoded previously into digital form, are transmitted together over high bandwidth systems. In such systems metallic wire conductors are gradually being replaced by optical fibres, augmented by broadband satellite transmission, enabling a radio or light beam carrier to replace the original much lower-frequency transmission methods.

Notwithstanding the introduction of special 'end-to-end' digital data networks carrying these services, the economics of existing PSTN and private analogue circuits used for data transmission prevent any very rapid conversion of the entire network to completely digital methods. At present the percentage take-up of the newer series represents only about 10% of the total possible network terminations (terminals and computers) and this figure is not expected to rise to more than about 40% by the end of the century. Also, for long-distance transmission of telephony signals, carrier systems using frequency division multiplexing will remain the norm throughout many parts of the world. This is due for the most part to the vast capital expenditure already made in existing carrier equipment.

Over the shorter trunk routes, however, digital transmission of speech telephony signals using **pulse coded modulation** (PCM) is gaining ground[1] but since these are exchange-to-exchange rather than end-to-end services they do not affect our conclusion concerning the primarily analogue nature of the present PSTN. What we are seeing within the various national telephone networks is essentially the creation of 'islands' of digital application surrounded by a 'sea' of analogue transmission equipment.

It is for these reasons that we need first to consider existing analogue carrier systems in some detail if we are to understand many contemporary computer communication systems and the standards under which they operate. In doing so we will also gain an insight into some of the problems that the newer all-digital systems set out to overcome.

Where possible in this chapter, we will use examples taken from the recommendations of the CCITT, the Comité Consultatif Internationalé de Télégraphie et de Téléphonie, an international body, not limited as the title suggests to the interconnection of telephone and telegraph circuits, but very much concerned with the establishment of data-carrying and network protocol standards. CCITT recommendations form a very important body of internationally agreed standards which have a profound influence on the way information transmission and particularly digital networks are developing. The recommendations themselves (note that these are recommendations — not standards) will be introduced as they occur in the text. A summary of the most important of these is given in an Appendix appearing at the end of this book. Further information can be obtained from the reports published by the International Telecommunications Union in Geneva (the parent organization for CCITT and itself part of the United Nations) and noted in various references given later. One which is particularly relevant to this chapter is given in ref [2].

2.1 THE PUBLIC SWITCHED TELEPHONE NETWORK

Speech or other analogue signals constituting a **communications channel** on the PSTN is only conveyed as far as the local **switching exchange** in its original 300 – 3400 Hz bandwidth form. At the exchange it is joined by many other channels which are combined together for onward transmission (switching) over a single wideband communications channel. The two processes which permit this to be carried out are **modulation** and **multiplexing**. A limited discussion of these principles is given below. Further information may be obtained from a number of standard texts that are available[3-6].

The process of modulation involves the variation of some parameter of one signal, **the carrier**, with another, **the message signal**. The modulated carrier can then be transmitted to its destination and an inverse process of **demodulation** invoked to recover the original modulation signal message.

The advantage of using a carrier system in the PSTN instead of transmitting the speech signal directly is that the speech signals are no longer limited to sharing a spectral location within the same low-frequency region of the spectrum. By suitable choice of carrier frequency we can allocate each separate speech signal its own region of the spectrum so that by making use of a series of adjacent regions, each having a bandwidth sufficient for a single speech channel, then the composite signal so formed will allow several speech frequency calls to be carried out simultaneously. We can see how this works by considering the simplest form of modulation, that of **amplitude modulation**.

2.1.1 Amplitude modulation

Amplitude modulation is illustrated in Fig. 2.1 for a carrier signal, $A_c \cos \omega_c t$ modulated by a message signal, $A_m \cos \omega_m t$, where $\omega_c = 2\pi f_c$, $\omega_m = 2\pi f_m$. Mathematically the process can be expressed as:

$$x(t) = (1 + m \cos \omega_m t) A_c \cos \omega_c t \tag{2.1}$$

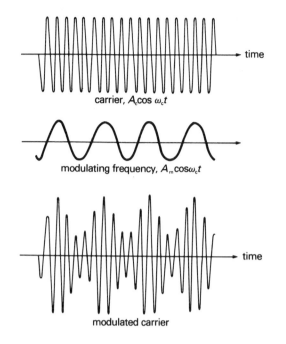

carrier, $A_c \cos \omega_c t$

modulating frequency, $A_m \cos \omega_c t$

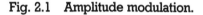

modulated carrier

Fig. 2.1 Amplitude modulation.

where m is known as the **modulation index** and represents a ratio of A_m/A_c which must be less than unity if distortion caused by signal amplitude limiting is to be avoided.

Expanding:

$$x(t) = A_c \cos \omega_c t + m[A_c \cos \omega_m t . \cos \omega_c t]$$

$$= A_c \cos \omega_c t + m/2[A_c \cos (\omega_c + \omega_m)t]$$

$$+ m/2[A_c \cos (\omega_c - \omega_m)t] \tag{2.2}$$

The resulting signal, $x(t)$, is seen to consist of three terms; a component at the original carrier frequency, f_c, and a pair of components at frequencies, $f_c + f_m$ and $f_c - f_m$. These latter components are known as **side-bands** and are the terms capable of carrying the message information.

Strictly speaking, only one of these terms is necessary to carry the modulation information along the transmission path. The carrier itself does not contribute to the intelligence transmitted and may be neglected, although synchronization difficulties may lead to this being transmitted as well but at a lower amplitude level. It is common practice to transmit only the lower side-band, when the procedure is termed **single side-band**

suppressed carrier modulation (SSB-SC). Thus no more bandwidth is required for this form of transmission than in the original message signal.

A disadvantage of complete carrier suppression is that the presence of the carrier is useful to permit accurate synchronization of the message signal. When the carrier is conveying digital data for example, the receiver needs to know the starting point of each bit in order to interpret the data correctly. A constant carrier provides a clocking mechanism to enable this to be carried out. A compromise method is **vestigial side-band** (VSB) transmission in which a reduced-power carrier and one side-band is transmitted.

Amplitude modulation represents only one of the possible methods of containing the message signal in the modulated carrier. We can also modulate the **frequency** of the carrier or its **phase** to convey the message signal. These will be considered later as a mechanism for transmitting binary information. First the process of multi-channel communication of amplitude modulation signals is considered.

2.1.2 Multiplexing

Amplitude modulation is applied in a **frequency division multiplexing** (FDM) context to combine many channels for transmission along a single path. The principle is shown in Fig. 2.2. Channel 1 is modulated by a carrier at frequency f_1. This results in two side-bands being produced as shown earlier,

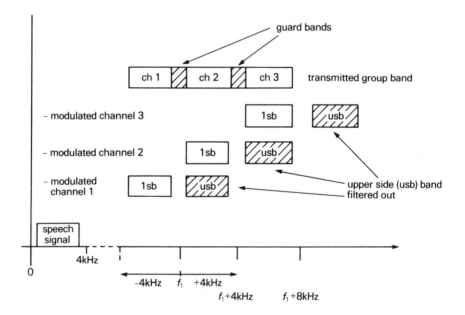

Fig. 2.2   Frequency division multiplexing in the PSDN.

and since speech is contained in the first 4000 Hz, the lower side-band is accommodated in the frequency range extending from $(f_1 - 4000)$ Hz up to f_1 Hz by the selective action of a band-pass filter. A carrier of frequency f_2 Hz which modulates the second channel is made equal to $(f_1 + 4000)$ Hz with the third channel carrier equal to $(f_2 + 8000)$ Hz and so on. The lower side-bands of these channels are selected by appropriate band-pass filters and mixed to produce the composite FDM carrier signal, which is transmitted for demodulation and separation into individual channels at the receiving end. The frequency range of 0 – 4 kHz is taken for each speech channel, rather than 300 – 3400 Hz, in order to prevent the separate modulated channels from overlapping in frequency. This 'guard-band' of 900 Hz between channels is useful also in that it accommodates the sloping sides of the selective filters used in a practical design.

Whilst in a **two-wire system** the telephone connection can enable speech to be carried in both directions simultaneously (**duplex working**), different arrangements are necessary in an FDM carrier system. This is because the system incorporates amplifiers which are unidirectional. Instead, a **four-wire system** is used in which signals for each direction of transmission are carried by a separate pair of wires. We will meet this distinction later when modem design for FDM systems is considered.

In the application of FDM principles to the PSTN, several internationally agreed groupings of channels occur. Twelve channels are combined as a basic **group** having a bandwidth of 48 kHz in the range of 60 – 108 kHz. The next basic building block is the 60-channel **supergroup**, formed by frequency division multiplexing five group signals. The subcarriers produced have frequencies from 420 to 612 kHz in increments of 48 kHz. The resulting signal occupies a range of 312 – 552 kHz. The next level in the hierarchy is the **mastergroup** which combines ten supergroups to form a 300-channel group occupying a range of 812 – 2044 kHz — a bandwidth of 1.232 MHz. Finally the **supermastergroup** combines 900 channels into a single group. These groupings form the CCITT international FDM carrier standards and are shown in Table 2.1.

Slightly different arrangements for the higher groupings are defined in the USA by A T & T.

Note that the original speech or data signal may be modulated many times and that the method of modulation may be different at different sections of the transmission route. For example, a data signal may be encoded using phase modulation to form a signal suitable for transmission over the PSTN. This, in turn, will be subject to several stages of amplitude modulation in the hierarchical FDM transmission system, and may also, in its passage through one of the digital 'islands' contained within the PSTN, be subject to PCM before finally reaching its destination. Each stage can distort the original data and contribute to the bandwidth/distance limitations of the total transmission system.

Table 2.1

CCITT international FDM carrier groupings

No of Channels	Bandwidth (kHz)	Range	Designation
12	48	60–108 kHz	Group
60	240	312–552 kHz	Supergroup
300	1232	812–2044 kHz	Mastergroup
900	3872	8.516–12.388 MHz	Supermastergroup

2.2 DATA ON THE TELEPHONE NETWORK

The circuit used to translate digital information into a form suited to voice-band transmission is known as a **modulator** and the circuit to perform the reverse function is known as a **demodulator**. Since transmission is normally **duplex** (i.e. in both directions at once) both circuits are required and the combination is called a **modem** (modulator/demodulator).

Early modems were expensive and limited in performance to a low transmission rate of 300 bps (full duplex) or 1200 bps (half duplex) over a normal two-wire PSTN 'dial-up' connection (i.e. one using the public telephone network rather than a leased line). By using alternative methods of modulation and particular combinations of these, it is now possible to choose signal conditions that give a much greater margin of tolerance to perturbation before the onset of errors and thus achieve considerably faster transmission rates. This accounts for the improved performance of modern modems and explains why the traffic rate has increased continuously in recent years using the same basic transmission media. In addition, the incorporation of micro-processor control has now enabled the modem to include quite sophisticated conversion, control and monitoring capabilities. An outline of these developments will be given below, commencing with a study of the different modulation techniques that may be used.

2.3 MODULATION METHODS

Three basic encoding techniques are available. These result in the modulation of the amplitude, frequency or phase of the carrier viz:

a amplitude modulation (AM),
b frequency modulation (FM) and
c phase modulation (PM).

These are general techniques applied to either continuous (analogue) or discrete (digital) modulation of the carrier. In the following, we consider their use as encoding methods for digital signals, compared in Fig. 2.3 using a rectangular modulating waveform. Applied to modem design they have all been used for different transmission standards either separately or in combination, when the process is termed **hybrid modulation** (HM).

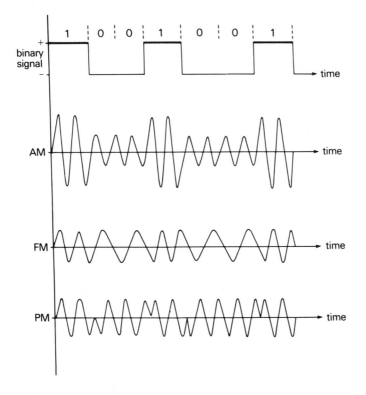

Fig. 2.3 Modulation methods for digital signals.

2.3.1 Amplitude-shift keying

The simplest of these methods is AM which we met earlier. Here the level of a single-frequency audio tone or carrier frequency, f_c, within the bandwidth possible by the speech transmission circuit, is switched or **keyed** between two levels designated as a digital 0 or 1 at a rate determined by the transmitted binary signal. This is known as **amplitude-shift keying** (ASK).

In ASK the two binary values are represented by two different amplitudes of the carrier frequency. The simplest method is to make one of these zero, that is, the absence of a carrier. Thus:

$$x(t) = \begin{cases} A \cos (2\pi f_c t + \theta_c) & \text{for binary 1} \\ 0 & \text{for binary 0} \end{cases} \tag{2.3}$$

where there is $A \cos(j2\pi f_c t + \theta_c)$.

The process generates a number of additional frequencies, the **side-bands**, and as in speech telephony an improvement in bandwidth efficiency is obtained by transmitting one of these side-bands and reducing (but not completely removing) the carrier to result in a vestigial side-band (VSB) transmission. This is effectively SSB-SC operation with the carrier restored to a small fraction of its original amplitude value.

These types of amplitude modulation are, however, subject to variability in transmission attenuation and have been replaced for modem operation by one of the more efficient methods described below. An exception is where AM is used in combination with other forms of modulation to achieve a high conversion efficiency. These forms of **hybrid modulation** will be considered later.

2.3.2 Frequency-shift keying

In **frequency-shift keying** (FSK) the binary digital 0 and 1 are represented by one of a pair of tones f_a and f_b. The two binary values are represented by two different frequencies near the carrier frequency, viz:

$$x(t) = \begin{cases} A \cos(2\pi f_1 t + \theta_c) & \text{for binary 1} \\ A \cos(2\pi f_2 t + \theta_c) & \text{for binary 0} \end{cases} \tag{2.4}$$

where f_1 and f_2 are offset from the carrier, f_c, by equal and opposite amounts.

Frequency modulation results in the generation of very many more side-bands than with amplitude modulation, and hence occupies a greater bandwidth. With FSK operating at very low bit rates the two side-bands can be considered as occupying two correspondingly narrow spectral lines, but as the bit rate increases, the frequency spectrum occupied spreads out on either side of the mean frequency. However, the method has the advantage of a constant power level and has a better signal-to-noise performance than ASK. It is fairly easy to generate and detect FSK signals and the method has enjoyed wide use in practical modem design for bit rates up to 1200 bps.

Full duplex operation over a two-wire telephone line is generally achieved by dividing the available bandwidth into two halves and using half for transmission in one direction and the other half for the reverse direction. The convention is always to use the higher frequency state for transmitting a binary zero. The CCITT frequency assignment for the 300 bps V21 recommendation using a two-wire switched circuit is given in Table 2.2. The incorporation of a modem into a data transmission system will be considered in Chapter 4, when interfacing procedures are discussed.

Table 2.2

Binary state	Forward direction frequency (Hz)	Reverse direction frequency (Hz)
0	1180	1850
1	980	1650

2.3.3 Phase-shift keying

Phase-shift keying (PSK) has proved to be a very successful form of modulation for digital transmission. The data signal is used to change the phase of the carrier frequency, f_c, by a fixed amount so that binary 0 and binary 1 can correspond to different phase shifts of the carrier signal.

In PSK the phase of the carrier signal, Θ_c, is shifted to represent binary 1 typically by $180° = \pi$ radians. Viz:

$$x(t) = \begin{cases} A \cos(2\pi f_c t + \pi) & \text{for binary 1} \\ A \cos(2\pi f_c t) & \text{for binary 0} \end{cases} \qquad (2.5)$$

A number of variations on this general theme have been developed. It has been found, for example, that PSK is susceptible to random phase changes and an improved method of phase modulation is to apply the shifts in carrier phase for each bit transmitted dependent on the logical state of the next bit to be sent. Thus a phase shift of 90° relative to the current signal indicates a binary 0 is being transmitted, whilst a phase shift of 270° would indicate a binary 1. Detection is then a matter of determining relative phase changes in consecutive samples of the carrier, a procedure which is easier to carry out. We can arrange this, for example, by deriving a local reference signal from the carrier and decoding the phase-shifted signal directly against this reference. This method is known as **differential PSK** (DPSK).

2.3.4 Multi-level signalling

In the examples discussed above, the bit rate is made equal to the signalling rate so that the bps and bauds are the same. In **multi-level signalling** it is possible to utilize more than two values of amplitude, frequency or phase so that each signal element may contain two or more bits of encoded information giving a bit rate greater than the symbol rate (bauds).

An example is the four-phase DPSK signal used in the CCITT V26bis data transmission recommendation that permits modem operation up to 2400 bps. Here, four different phase changes are employed (45°, 135°, 225° and 315°) instead of just two, so that the phase change of each signal can convey two bits of information (Fig. 2.4). This scheme is often known as **quadrature phase shift keying** (QPSK), viz:

$$x(t) = \begin{cases} A \cos(2\pi f_c t + 45°) & \text{for binary } 00 \\ A \cos(2\pi f_c t + 135°) & \text{for binary } 01 \\ A \cos(2\pi f_c t + 225°) & \text{for binary } 11 \\ A \cos(2\pi f_c t + 315°) & \text{for binary } 10 \end{cases} \qquad (2.6)$$

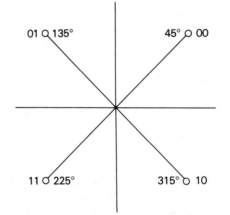

Fig. 2.4   Quadrature phase-shift keying

In order to reduce coding errors in transmission a **gray coding** is applied (Table 2.3) in which only one bit changes from each state to the next adjacent state. Whilst this scheme has been extended to eight phases (CCITT recommendation V27), increasing transmission rate to 4800 bps, further increase is limited by the bandwidth and level of noise on the line.

There is, in fact, a direct relationship between the maximum information rate, bandwidth and noise, which was first derived by Shannon[7]. This is the **Shannon–Hartley Law** and is expressed in terms of a maximum theoretical channel capacity, C as:

Table 2.3

Gray coding	
Decimal	Code
0	0000
1	0001
2	0011
3	0010
4	0110
5	0111
6	0101
7	0100
8	1100
9	1101
10	1111

$$C = B \log_2(1 + S) \quad \text{bps} \tag{2.7}$$

where B is the channel bandwidth (Hz) and S is the signal-to-noise power ratio (SNR) (expressed as a ratio and *not* in dB).

A typical telephone channel could have a bandwidth of about 2700 Hz and an SNR of at least 30 dB (i.e. a power ratio of 71,000). This results in a theoretical maximum control capacity of $C = 27$ kbps. It should be stressed, however, that this is a maximum rate. Less than half of this is realized due to cross-talk, phase jitter and other line impairments. Thus, 9600 bps is about the maximum possible with a good PSTN leased line. Nevertheless, the principle of multi-level coding is seen to offer a considerable improvement in bandwidth efficiency over earlier methods. A similar technique can, of course, be applied to AM using VSB transmission, but no advantage is gained over DPSK, whilst the modem implementation becomes more difficult. (There are no CCITT recommendations for AM–VSB in speech-band modems.)

2.3.5 Hybrid modulation

Although further improvement in bandwidth efficiency by extending DPSK to 16 or more states is ruled out by increased susceptibility to noise etc., it is still possible to realize greater bandwidth efficiency by using a combination of amplitude and phase modulation. This is achieved by exploiting the fact that the DPSK modulation is equivalent to two separate baseband signals transmitted independently on a carrier but with a 90° phase difference; i.e. simultaneously using the In-phase and Quadrature components of the same carrier, f_c. The two signals are then combined and transmitted to the line as a composite signal. Decoding is obtained by demodulating each channel by an appropriate in-phase and quadrature component of the recovered carrier[8].

This is illustrated in Fig. 2.5 where an incoming signal is presented as two separate streams, each applied to the I and Q channels as described above. By independently modulating the two signals, using AM having four amplitude levels, when these are combined for transmission 16 possible phase/amplitude states are present, thus permitting each transmitted symbol to represent a group of four digits.

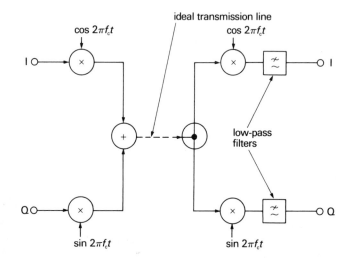

Fig. 2.5 Quadrature amplitude modulation.

This method is termed **quadrature amplitude modulation** (QAM) and is illustrated in Fig. 2.6a by means of a **signal-space diagram**. This is the signal constellation used for the CCITT V29 9600 bps recommendation. The circles indicate the maximum limits of noise/jitter permissible at each phase/amplitude level before an error can occur. A slightly different combination of 16 phase/amplitude levels is the 'four-square' structure shown in Fig. 2.6b which produces a rather better noise performance. This is applied in CCITT recommendations V22bis (2400 bps) and V32 (9600 bps). Higher rates are possible under certain leased-line conditions, and a recommendation (V33) is made for a 14400 bps system. This appears to represent an upper limit in the exploitation of carrier-borne digital transmission due to deteriorating SNR and susceptibility to phase/amplitude errors that occur when a greater number of levels is attempted.

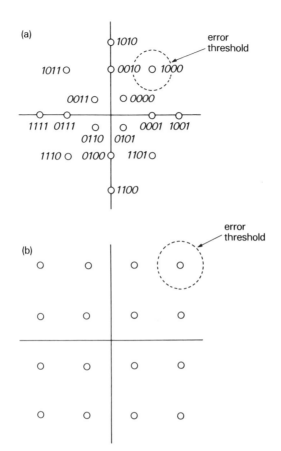

Fig. 2.6   Signal-space diagrams for QAM.

2.4   MODEM CATEGORIES

With increase in transmission speed comes greater complexity in modem design, as we have just seen. In terms of speed of operation we can categorize modulation methods as:

a   lower speed (up to 300 bps) using FSK;

b   medium speed (up to 1200 bps) using DPSK; and

c   high speed (2400 bps or more) using QAM.

Each of these speeds relates to one or more CCITT recommendations[9], which not only define the modulation scheme employed but also the key parameters which must be met to ensure compatible operation between similar modems. A summary of these recommendations is given in Table 2.4.

Table 2.4

CCITT recommendation			
CCITT recommendation	Line speed (bps)	Method	Comments
V21	300	FSK	2-wire duplex
V22	1200	PSK	2-wire duplex
V21bis	2400	QAM	2-wire duplex
V23	1200/75	FSK	2-wire duplex at 1200/75 half duplex at 1200/1200
V26, V26bis	2400	PSK	half-duplex
V26ter	2400	PSK	half-duplex (echo canc.)
V27, V27bis, ter	4800	PSK	half-duplex
V29	9600	QAM	half-duplex
V32	9600	QAM	2-wire duplex (echo canc.)
V33	14400	QAM	4-wire leased circuits

The V23 recommendation is unusual in that two line speeds are specified for duplex working. One of these relates to the line speed in one direction and the other for the reverse direction. This would be used, for example, in an enquiry service, such as the British Telecom Prestel Service, where the lower speed is used for the interrogation (i.e. from the keyboard) and the higher speed for the greater volume response data (i.e. picture display signal).

In order to facilitate the transfer of large programs and data to personal computers via a modem, recommendation V32 is proposed. This can operate at 9.6 kbps in a 'dial-up' situation and is a considerable improvement on the rate of 1200 bps or less used with earlier equipment.

Further CCITT recommendations which have general relevance to data transmission and hence modem operation are considered in Chapter 5 when digital interface standards are discussed.

2.4.1 Baseband modems

Whilst not strictly modulation/demodulation systems, a number of devices are available which perform similar functions but operate directly on the digital input signal. These are referred to as **baseband modems** or **line drivers** and carry out data transmission without modulating or demodulating a carrier signal. The digital signal is first filtered to reduce the high-frequency content and the modified signal is transmitted directly over the line media.

Baseband modems are only suitable for transmission over short distances but can achieve data rates up to 19.2 kbps.

2.4.2 The codec

Whereas the modem enables digital data to be represented by an analogue carrier signal, the **codec** (coder/decoder) allows analogue data to be represented by digital signals. This would used, for example, in transmitting and storing speech information in digital form over a data network. The analogue signal is subject to an analogue/digital conversion process to approximate the signal to an equivalent bit stream. At the receiving end the bit stream is used to reconstitute the analogue signal. In operation it has been found that the digital data transmitted often contains strings of repeated characters. This can result in spectral peaks in the transmitted information which can cause interference with adjacent channels. To overcome this, the codec randomizes the data by scrambling the bit stream using a shift register and uses a complementary de-scrambler at the receiving end. The scrambling polynomial forms part of the appropriate CCITT recommendation. This effect can, of course, occur with any data transmitting device so that a scrambling routine is commonly found in data transmitting equipment.

2.5 OTHER MODEM FACILITIES

The modern modem does more than simply convert digitally coded information into a form suitable for transmission over a telephone network. Frequently the modem is designed for **duplex** operation, that is, the simultaneous transmission of data in both directions over two-wire telephone circuits. The most common method for bit speeds of up to 1200 bps is simple frequency division multiplexing, noted in section 2.1.2, in which the telephone channel divided into two separate 'go' and 'return' frequency channels. For higher speeds, a system of **echo cancelling** is used in which the full telephone bandwidth is employed for both directions of transmission, and adaptive cancelling circuits are used in each receiver to cancel the interference from both its own transmitter and its returned echoes[10].

In addition, the modem is likely to contain other electronics such as equalization filters to minimize channel distortion (discussed in the next chapter), scrambling and unscrambling devices to randomize the data input, automatic answering facilities, automatic channel speed recognition and bit rate switching capability (the so-called 'frequency-agile' modems). A recent development is the incorporation of automatic line/modem testing and diagnostic facilities. These take the form of stored test patterns which are transmitted to the distant modem via the transmission line. A loop configuration returns the signal for analysis by the modem electronics. The basic requirements for these procedures are specified in a further CCITT recommendation, V54[5].

The incorporation of a microprocessor within the modem design has enabled modem manufacturers to incorporate these and other 'intelligent' control features at very little extra cost. A most important member of these additional facilities is the provision of a multiplexing capability which we discuss below.

2.6 MULTIPLEXING

Applying a data network to establish connection with a large number of terminals or devices requires multiple channels and these need to be transmitted as a group if the costs of communication are to be kept low. A solution is to use multiplexing techniques either in the form of a computer network, which can be considered as a form of distributed multiplexer sharing communication time or bandwidth between connected devices, or as a hardware multiplexing unit or concentrator to combine the data traffic from a number of devices into a single stream for transmission over the medium. We consider the network solution in some detail in later chapters. Here the development and function of multiplexer and concentrator techniques will be discussed.

The way these two devices function is somewhat similar. A **multiplexer** attempts to maintain data transparency, in the sense that the data input is identical in form to the data output although it will be combined with other data from other channels. A **concentrator**, on the other hand, can operate on the data as well, thus carrying out some data processing on the signal(s). An example would be the removal of blank spaces included in the input data, thus applying a form of **code compression** to the message and so reducing the transmission bandwidth requirements. Sufficient information would be appended to the transmission to enable a corresponding insertion of missing blank spaces, i.e. **code expansion**, at the receiving end.

The two basic alternatives in multiplexer operation are:

a **frequency division multiplexing** (FDM) in which the available signal bandwidth is divided into separate sections with each section used for one channel of communication; and

b **time division multiplexing** (TDM) in which the time available for transmission is subdivided into separate time slots, with each slot used for one channel of communication.

2.6.1 Broadband and baseband

The use of FDM to share the circuit capacity of a single transmission line is called **broadband signalling**. When broadband systems are used for data transmission a separate modem operating frequency is required for each channel and this frequently needs to be different for each direction of transmission. Thus the modems used in this mode form a complementary pair, the transmitting channel used by one modem becomes the receiving channel of the other and vice versa. When large numbers of channels are involved the number of modems needed becomes considerable.

Two solutions to this problem are applied. One is to employ **frequency-agile modems**. These are modems which are able to change operating frequency in response to a signal sent from a central controlling device or computer. The second is to route all the transmitted signals to a single device, called a **headend** which converts a whole set of transmitted frequencies into another set of frequencies used only for reception. The advantage is that the pairs of modems need not be complementary but can be identical, both transmitting on one frequency and receiving on the other. This method finds wide use in broadband local area networks which are discussed in Chapter 7.

Where digital data are conveyed directly over the media, the entire capacity of the line is used to support the transmission of a single 0 or 1 bit at a time. This is called **baseband signalling** and applies where only one digital signal is transmitted, as in point-to-point connection, or in a TDM situation where the *time* available for transmission of a number of signals is shared. At any given time only one digital signal element can be present on the transmission line.

Two basic problems are associated with baseband transmission. First, only one device connected to the line can transmit at a time since otherwise the transmitted data will be corrupted. Second, it is necessary to know precisely the time and duration of each digit transmitted so that their logical value can be recognized at the receiving end. Additionally, in the case of TDM the channel allocation for each of the transmitted digits is required to be known. The first implies that means must be found to recognize when two or more devices attempt to transmit at the same time and to deal with the consequences should it occur. It is a difficulty with broadcast media such as local area networks and fairly complex methods are needed to overcome it. The second is a problem common to any direct digital coding method and will be considered in Chapter 4 when coding methods are discussed. Methods of channel synchronizing for TDM are described in section 2.6.3.

2.6.2 Frequency division multiplexing

This is possible when the useful bandwidth of the medium exceeds the required bandwidth of the signal to be transmitted. It is essentially an analogue method and was discussed earlier in section 2.1.2.

Despite the continuing widespread use of FDM as a means of accommodating a large number of speech-grade channels over a wideband PSTN, it has not been used to any extent for the transmission of multi-channel digital data following the introduction of TDM techniques for this purpose in the 1960s. This is due to a number of problems that can arise with FDM such as **cross-talk**, when the spectra of adjacent channels will interfere with one another and may overlap (although unmodulated **guard-bands** can be situated between channels to minimize this), and **intermodulation noise** resulting from the non-linearity of the transmitting equipment. FDM is also expensive to install and maintain in comparison with TDM.

2.6.3 Time division multiplexing

Time division multiplexing is applied in two forms:

a **synchronous TDM** commonly used for multiplexing digitized speech signals, and

b **asynchronous TDM** which is also known as **statistical TDM** for reasons which will be apparent later.

With both forms of TDM, each channel is allocated the full digital bandwidth of the communication link in turn, giving to each terminal user the impression of realizing a fraction of the total bandwidth, but allocated permanently. This gives rise to a difficulty that a time-slot allocation is made repeatedly for a given channel whether this has or has not data to transmit at the time of allocation. Due to the dialogue or 'bursty' nature of much computer generated data, this can result in idle channels with consequent waste of available bandwidth. The incorporation of a controlling microprocessor within the TDM or shared in the associated modem circuit has permitted the development of the **statistical TDM**, or 'statmux' to overcome this. The statmux operates on a similar time-slot allocation basis but here the slot is made variable in length, requiring only a flag bit to be transmitted if no data are present.

These three alternative types of multiplexing operation are compared in Fig. 2.7. Note that in the statmux example channels, 1, 2, 4 etc. are narrow and could represent channels each containing a single flag bit, permitting other channels, e.g. 3, 5, 6 etc. to be expanded to carry more information.

2.6.4 Choice of multiplexing technique

TDM can be used to interleave bits or characters from various channels being multiplexed. **Bit-interleaved multiplexers** are used with a synchronous data

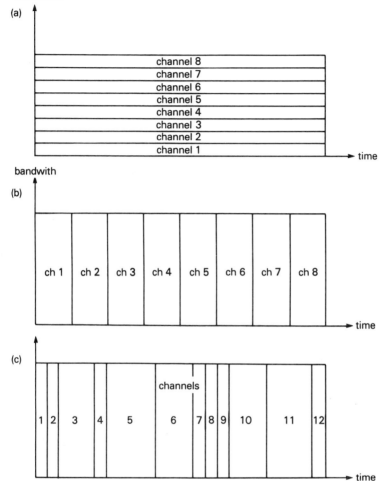

Fig. 2.7 Multiplexer operation. (a) Frequency division multiplexing; (b) time division multiplexing; (c) statistical TDM.

source. Each time slot contains just one bit. They are fast in operation and, since very little delay is experienced, are suitable for medium- to high-speed devices such as synchronous VDU terminals. A single bit buffer is included in each incoming data circuit. Character devices operating under start–stop control require a **character-interleaved multiplexer**. A common buffer holding characters for all channels may be included to make better use of the single high-speed communications channel.

With a synchronous TDM, i.e. one allocating a time slot to each channel in strict rotation, if all the channels operate at the same speed then there is no need to identify a particular channel since they all follow in a recognized sequence. It is, however, necessary to maintain overall synchronization between transmitter and receiver since, if transmitter and receiver are out of step, then data on all channels are lost. To achieve this the data are organized into frames with a synchronizing digit added, as shown in Fig. 2.8a.

(a)

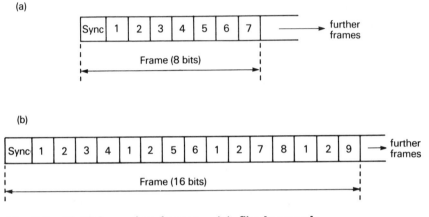

(b)

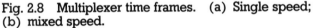

Fig. 2.8 Multiplexer time frames. (a) Single speed; (b) mixed speed.

A **frame** consists of a group or block of binary digits, generally including additional bits describing the contents of the frame in some way. Transmission consists of sending a sequence of consecutive frames across the network which can be individually checked at the receiving end. Further implications of frame transmissions will be described in Chapter 4.

A common TDM synchronizing mechanism is to add one control bit to each TDM frame. This is known as **added-digit framing**. These can be alternate 0s and 1s to give a frame/frame pattern of 0101010101... which is unlikely to occur continuously in the data. The receiver can search for this pattern and, once found, check that it is maintained and hence correct synchronization realized.

Since the data for each channel come from a different source, each with its own local clock, there is a possibility that the rates may differ by other than a rational number and hence throw the individual bits in the frame out of synchronization. A simple technique established to overcome this is known as **pulse stuffing**. (This is different to **bit stuffing** which we come across later when considering protocols and error checking.) The outgoing data rate of the multiplexer is made slightly higher than the sum of the maximum instantaneous incoming rates. The extra bit capacity is made up by inserting additional pulses at fixed locations in the frame format so that they may be identified on reception and removed.

It is possible to handle mixed speeds with this type of **fixed frame** multiplexing by permitting the faster speed to use more than one slot in a frame period. The constitution of the 'mix' must be known to both ends, however, and any unused slots filled with padding characters. An example is shown in Fig. 2.8b, in which seven channels of 300 bps and two at 1200 bps are combined for transmission over a 4800 bps link. In contrast, the single-speed frame shown in Fig. 2.8a could represent the result of combining seven channels at 1200 bps into a 9600 bps link (one additional channel is required for synchronization).

If the padding characters could be removed and replaced by 'channel absent' bit flags then a considerable gain in efficiency could be realized. This is the principle of the statmux, which typically offers a throughput factor of about 2 – 4 times compared with a normal TDM[11].

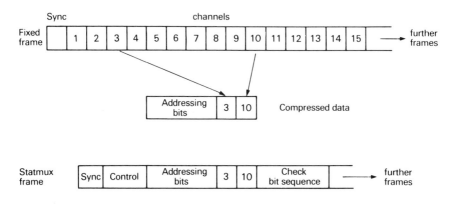

Fig. 2.9 Comparison between fixed frame and statistical multiplexer frame.

The frame layout for statmux operation is shown in Fig. 2.9 where it is compared with fixed frame TDM. Assuming that only two channels, numbers 3 and 10 in the fixed frame, contain actual data, with padding bits only present in the other channels then, as shown in the compressed data frame, only the data from these channels need be included. Some addressing bits are added to indicate the position of the data channels in the original polling sequence. In statmux operation it is usual also to include some form of data checking so that the transmitted information can be checked for errors upon reception. (We will be considering how this checking is carried out in a later chapter.) In many sophisticated statmux designs a control field is included, as shown at the front of the data field, where it is used to insert acknowledgements and retransmission requests.

Note that the statmux dynamically allocates time slots on demand. At the input to the multiplexer, the input buffers are scanned and data collected where present, until a complete frame output buffer is filled. The frame is then transmitted together with address information. At the receiving end the multiplexer receives a frame and distributes the data to its appropriate output buffers in accordance with the address information contained within the transmitted frame. Thus the actual data speed in bits per second is irrelevant and special arrangements to permit additional time slots for faster speed data are not required.

SUMMARY
Chapter 2

This chapter is concerned with data transmission over an analogue carrier system. Despite the emergence of full digital transmission systems, the use of modems to permit the public switched telephone network (PSTN) to be used for this purpose is likely to be the dominant method of data transmission for some time to come.

The different modulation methods applied to data transmission using modems are amplitude-shift keying, frequency-shift keying, phase-shift keying and differential phase-shift keying. More complex methods of multi-level signalling and hybrid modulation are used to obtain higher bandwidth utilization for modem operation, of which quadrature amplitude modulation represents a major advance. A series of CCITT international procedural recommendations are applied to these methods to permit the exchange of data between different end systems.

Application of basic modulation methods in modem design is assisted by the incorporation of a controlling microprocessor within the modem, which is also used for a number of ancillary tasks, including multiplexing. Multiplexing of a number of channels into one transmission route is achieved by frequency division multiplexing (FDM) or time division multiplexing (TDM). The former is used mainly in the PSTN whilst the latter is applied to data transmission. Alternative methods of TDM are operation with a fixed data frame or with a variable-length data frame. Variable length data frames are used in the statistical multiplexer which is shown to be more economical in use of the available bandwidth than fixed-length frame multiplexing.

**PROBLEMS
Chapter 2**

P2.1 **a** Assuming a *carrier frequency* of $A_c \cos \omega_c t$ *and a modulating frequency of* $m \cos \omega_m t$ write down an expression for amplitude modulation and from this determine the ratio of side-band power to total power in terms of the modulation index, m.

b What is this ratio if the form of modulation is single side-band?

P2.2 Compare the susceptibility to error of the two forms of QAM configuration shown in Fig. 2.6. Give an approximate value in decibels showing their difference in performance.

P2.3 Draw an approximate frequency spectrum for a two-wire full duplex FSK system designed according to CCITT recommendation V21. Comment on the spectrum you have drawn.

P2.4 A method of modulation similar to SSB-SC modulation is often used in which the carrier is not completely suppressed but reduced to a small power level. This is known as vestigial side-band (VSB) modulation. Suggest reasons for adopting this form of modulation in a practical situation.

P2.5 Discuss the techniques of time division multiplexing and describe the operation and advantages of using statistical multiplexers in a PSTN carrier system using modems over a leased line. Show how the technique could be further improved by applying concentrator principles in the statmux.

P2.6 Give the difference, if any, between the demodulator part of a modem and the coder part of a codec. Do they do the same job?

3

DATA TRANSMISSION

Efficient data transmission is highly dependent on the characteristics of the transmission medium and on the correct matching of the selected medium with the form of the transmitted signal. In this chapter, the common types of transmission media are discussed and their effect on signals being conveyed along them. The transmission medium is the physical path between transmitter and receiver. In many cases this takes the form of a pair of metallic conductors, but it can also be achieved through the transmission of a light beam along a glass fibre or by means of an electromagnetic beam through free space.

In any medium there are various impairments to the signal that can occur during the transmission process. These are, principally, distortion as a function of distance or frequency, electrical noise and signal time delay. The effect of these impairments is to cause errors in data transmitted through the medium. Many of the techniques for signal modification and encoding before transmission are directed towards reduction of these errors and detection of those that do occur. Some of the basic ideas of signal modification to minimize the transmission of errors will be discussed towards the end of this chapter.

3.1 TRANSMISSION MEDIA

3.1.1 Twisted pair

The simplest of these transmission media consists of a pair of conductors, an **open line**, which, in the case of inter-equipment communication, often takes the form of a flat ribbon cable, containing several lines and a common return wire. This is, however, extremely susceptible to inductive- or capacitative-induced noise signals from neighbouring lines or external radiating sources (e.g. power lines) and much better **noise immunity** is obtained by the use of pairs of wires which are twisted together. The close proximity of the **twisted pair** lines to each other means that any interference is induced in both wires simultaneously and is of equal strength so that its effect on the difference signal existing between the pair of wires is reduced. The twisted pair finds wide use in the PSDN and in local area networks where the cable length is short, such as within a single building, although twisted pairs have been used successfully for direct digital transmission up to 15 km. The attenuation with such a cable is very frequency dependent and also depends on distance as shown in Fig.1.1. It is possible to improve the frequency characteristic by means of loading coils and whilst this is carried out for carrier telephone communication it is not practical for direct digital communication[1].

Compared with other transmission media, twisted pairs are extremely limited in length before the signal deteriorates to an unusable level. When carrying analogue signals it is necessary to insert **amplifiers** to boost the signal every 5 or 6 km. These are devices having a linear input/ouput characteristics to faithfully reproduce at the output an amplified version of the input signal. With digital data **repeaters** are required for each 2 or 3 km. A repeater differs from an amplifier in that the incoming signal is amplified (and the amplification need not be linear) and then attenuated to provide a copy of the input pulse but having a more rectangular waveform characteristic.

Although reduced by twisting the two conductors together, some inductive effects remain, particularly from adjacent twisted pairs carrying other signals. This is known as **cross-talk** and is a major problem with analogue or carrier-borne systems. Twisted pair cables are limited in bit rate transmission particularly over long distances and a data rate of a few million bits per second represents a reasonable upper limit.

3.1.2 Coaxial cable

A much better performance is obtained through the use of **coaxial cable** (Fig. 3.1.) This consists of a hollow outer cylindrical conductor which surrounds a single inner wire conductor. The inner conductor is held in place either by regularly spaced insulating rings or by a solid dielectric material. The outer conductor is covered with an insulating sleeve. Greater bandwidth is obtained with coaxial cable compared with twisted pair, enabling digital transmission rates of up to 140 Mbps to be obtained over short distances. In order to achieve low transmission loss it is necessary to use large cable diameters. This arises because at high frequencies the current is concentrated near the surface of a conductor. Owing to this so-called **skin effect**, an increase in cable diameter is required as the operating frequency (i.e. bit rate) is increased if the effective cross-sectional area of the conductor is to be maintained. A coaxial

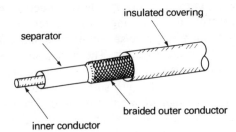

insulated covering

separator

inner conductor

braided outer conductor

Fig. 3.1 A coaxial cable.

structure is able to provide much reduced cross-talk compared with a twisted pair. This is because the electric and magnetic fields are almost entirely constrained within the cable, the outer surrounding conductor being earthed.

Coaxial cables can be used for a number of different signal types with transmission rates of up to 10 or 20 Mbps over several hundred metres. For local area network transmission two types of coaxial cable are currently in use: a 75 ohm cable which is also used as a standard in community television projects and a 50 ohm cable. These are sometimes referred to as **broadband** and **baseband** coaxial cables, respectively, for reasons which will become apparent later.

3.1.3 Optical fibres

Guided-wave optical communications systems using **optical fibre** are particularly suited to digital communication. The attenuation with distance is low, the transmission bandwidth is high and the cables are immune to induced electromagnetic fields[2]. The ability to transmit a light beam along a light conducting medium has been known for some time. What is new is the ability to modulate such a light source with a binary digital signal at very high bit rates and to extract the modulated signal at the receiver. This is achieved by the use of solid state light sources, the **light-emitting diode** (LED) or the **injector laser diode** (ILD), and a **photodiode** to convert light into pulses of electrical energy at the receiving end.

The advantages of optical fibre over other transmission systems may be summarized as:

a greater bandwidth and hence a potentially high transmission capacity;

b smaller cable size and lighter weight;

c lower attenuation over a greater distance;

d greater repeater spacing;

e negligible cross-talk;

f High immunity to interference; and

g complete electrical isolation.

An optical fibre cable consists of a glass or plastic core that is completely surrounded by a cladding material having a slightly lower refractive index. In effect, the fibre optic cable acts as a waveguide whereby the optical signal is guided along the glass core through being constrained by reflections at the core/cladding boundary.

The mechanism for the transmission of the encoded light beam is that of **total internal reflection**. Total internal reflection, illustrated in Fig. 3.2, can occur at the boundary with any transparent medium that has a higher index of refraction than the surrounding medium. Light from a source is directed into the core and those rays having a shallow angle less than Θ_1 will be reflected at the core/cladding boundary and propagated through angle Θ_2 along the fibre. Other rays at less than the critical angle, Θ_1, will be absorbed by the cladding material or passed out of the fibre altogether. The modulated light signal thus travels along the central core by means of a series of total internal reflections or zig-zag excursions as seen in the diagram.

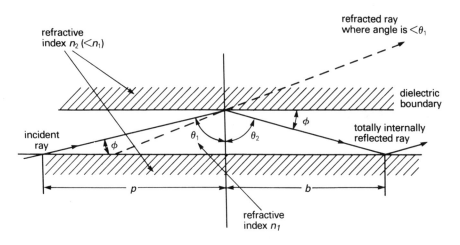

Fig. 3.2   Total internal reflection.

The simplest form of optical fibre has a relatively large central core (about 50 mc in diameter) having a refractive index n_1 and surrounded by a cladding material having a lower refractive index, n_2. This type of cable is known as a **stepped-index multimode** cable and is shown in Fig. 3.3a.

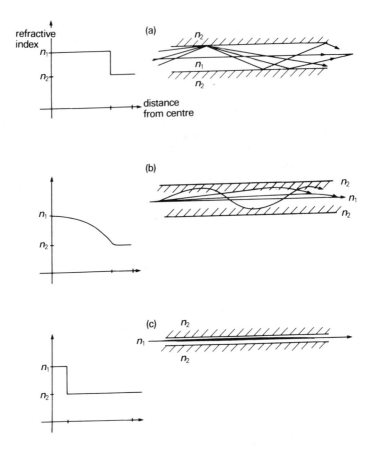

Fig. 3.3 Fibre optic cables. (a) Stepped-index multimode;
(b) graded-index multimode; (c) stepped index monomode.

A short pulse of light entering such a cable is broadened by the multiple reflections that occur at the boundaries. This is because the light is not constrained precisely to one particular angle relative to the core axis. There are, in fact, a number of possible discrete modes which result from insertion angles less than Θ_1 and these give rise to a series of N discrete propagation angles Θ_2. The total number of these modes, N, is proportional to the square of the inner core diameter and is significant for core diameters greater than about 10 mc.

For these large-core diameters, the actual delay is given by:

$$\text{delay period} = (t_2 - t_1) = \frac{n_1 \, L}{n_2 \, c} \, (n_1 - n_2) \tag{3.1}$$

where L is the cable length and c the velocity of light.

This pulse broadening effect is known as **multipath pulse dispersion** and limits the bandwidth of a multimode cable. A 50 mc diameter multimode cable generally operates in the 850–900 nm wavelength region when it is limited to bit rates of about 140 Mbps for an 8 km repeater spacing[2]. Faster operation and/or longer repeater spacings are achieved through other fibre optic cable designs which are considered next.

One method is to construct a fibre optic cable using a graded refractive index profile. Such a fibre does not have a distinct core and cladding boundary. Instead, the index changes progressively so that it is highest in the centre and then decreases towards the fibre boundary. This is known as a **graded-index multimode** cable and is shown in Fig. 3.3b.

An optimum index distribution corresponds to a near parabolic refractive index profile. With this type of profile many modes of propagation occur simultaneously. The modes travelling in the outer regions of the fibre are in a region of lower refractive index and therefore travel faster than those travelling in the higher refractive index near to the fibre axis. The net effect is that a more nearly equal delay is experienced between different path lengths and less pulse broadening occurs.

The pulse spread for a graded-index fibre is given approximately as:

$$\text{delay period} = t_2 - t_1 = \frac{L}{2c} n_{\max} d^2$$

$$(3.2)$$

where d is a parameter which determines the ratio of change of the refractive index (about 0.01) and $n_{\max}$ is the maximum refractive index.

Considerable reduction in pulse spreading is obtained for graded-index fibre, amounting to about 100 times in favourable cases. Thus the signal distortion of graded-index fibre is much less than with an equivalent stepped-index fibre, enabling much larger bandwidths to be obtained.

Although graded-index fibres offer better peformance, they too eventually present a bandwidth limitation for higher bit rates, and in addition carry a disadvantage of more expensive production costs. An alternative approach is to reduce the size of the central core to reduce multipath dispersion to near zero value. This is possible since the number of discrete propagation angles, Θ_2, decreases rapidly with very small core diameters. Ideally, we would like to use fibres so small in diameter, compared with the wavelength of the optical signal, that only one mode is possible so that the signal would go straight down the middle of the cable with no zig-zagging at all. This can occur with a core diameter of about 9 mc or less when the cables are referred to as **stepped-index monomode** cables (Fig. 3.3c). However, as the number of modes that can propagate along the core is reduced by reducing the diameter of the fibre core, two other dispersive effects begin to dominate. These are **material dispersion** and **waveguide dispersion**.

Material dispersion occurs since the refractive index of the medium, and hence propagation speed, varies with the wavelength of the transmitted light. The light energy of a LED, for example, comprises a number of different frequencies whose time to travel through the fibre will be different, resulting in a dispersed pulse at the far end.

Waveguide dispersion is the distortion of the optical signal arising from the dependence of the phase and group velocities of each mode on the wavelength and is a result of the geometric properties (lack of symmetry) in the optical fibre. Fortunately, material dispersion and waveguide dispersion act against each other with respect to frequency as we see from Fig. 3.4a. By choosing the operating wavelength carefully, a cancelling effect can be realized thus minimizing the effects of distortion. One wavelength at which this can be achieved is 1550 nm, and this also happens to coincide approximately with one of the regions of minimum spectral absorption for the fibre optic material itself (a second is at 1300 nm) (Fig. 3.4b). This has led to monomode fibres operating at 1550 and 1300 nm becoming the dominant standards in fibre optic technology and their adoption by, for example, the UK PSTN trunk service[3].

The use of such cables enables the distance between repeaters to be increased to at least 30 km. Current development is directed towards increasing this for such applications as submarine communication cables, where a reduction in the total number of repeaters can reduce installation and maintenance costs considerably. Some characteristics of an advanced mono-mode application showing the potential of this technique are reproduced in Table 3.1 (by permission of British Telecom).

3.1.4 Radio transmission

The transmission media so far discussed have all used a physical medium to carry the transmitted information. This can be termed the **hardware media**. Data can also be propagated using radio waves in free space, termed the **software media**. One example is the use of line-of-sight microwave radio transmission. This may be applied to overcome situations where cable cannot easily be laid, such as to link buildings on adjacent sites separated by a public road. Such **terrestrial links** are subject to electrical and atmospheric disturbance, a disadvantage which is not present with **satellite transmission** where the radio beam travels most of its path through free space.

Terrestrial microwave systems for long haul telecommunications are used as an alternative to coaxial cable or fibre optics links. They are also being considered for short point-to-point links as part of a local area network[4].

As with coaxial cable systems, the microwave link can support high data rates over long distance but will require fewer repeater stations over the same distance. Unlike other forms of transmission, the medium plays a much smaller part in determining the transmission characteristics. Instead, the

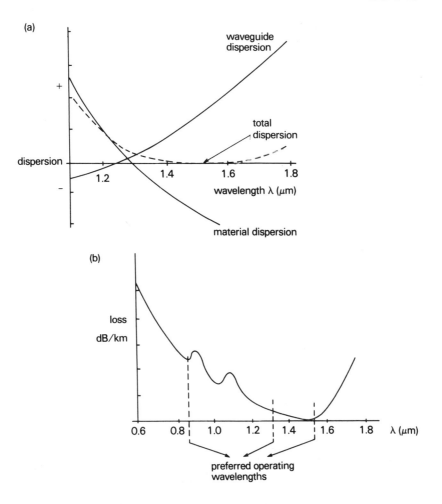

Fig. 3.4   (a) Dispersion effects with monomode cables.
(b) Operating wavelength for monomode cables.

Table 3.1

Characteristics of a type T fibre optic cable

	a	b	c
Bit rate (Mbps)	140	320	650
Link length (km)	104	104	83
Launch power (dBm)	−7.8	−7.8	−7.8
Loss (dB)	34	34	26
Received power (dBm)	−42	−42	−34

Included with permission of British Telecom.

important parameters are the design of transmitting system and the carrier frequencies involved. At UHF frequencies the radiation system consists of a parabolic dish which is highly directional. The directional characteristic is important since it is desirable to minimize the radiation power necessary to reach a given objective. Directivity is related to the ratio of the dish diameter, D, and the wavelength, λ, and is usually stated as a 'gain', G, in a particular direction compared with that of an omnidirectional aerial and is given approximately by:

$$G = (D/\lambda)^2 \text{ dB} \qquad (3.3)$$

The range, d, between communicating line-of-sight parabolic aerials with no intervening obstacles is given by:

$$d = \frac{7.14}{K h} \text{ km} \qquad (3.4)$$

where h is the aerial height (km) and k is a factor take into account the curvature of the earth (approx. 1.33).

The attenuation, as with any form of radio propagation, varies with the square of the distance from the transmitter. Common frequencies used for transmission are in the range 2 – 40 GHz when multiplexed frequency modulation carrier systems are used to convey the data transmitted.

A communication satellite is, in effect, a microwave relay station. It is used as a device for point-to-point and broadcast transmission[5]. The satellite receives a transmission on a frequency band known as the **uplink** and retransmits down to the ground station on another frequency — the **downlink**. The frequencies used ideally should be between 1 and 10 GHz and, although attenuation by atmospheric absorption is greater above 10 GHz, higher frequencies than this are beginning to be used due to saturation of the ideal band range. The INTELSAT series of geostationary communications satellites generally use 6 GHz *up* to the satellite and 4 GHz down to earth. The effective transmission bandwidth is 80 MHz for the earlier satellites and up to 2300 MHz for more recent ones.

The lengthy transmission path will result in a considerable transmission signal delay of about 240 – 300 mS. This is noticeable in ordinary telephone conversation and can introduce problems in digital data transmission, particularly in the correction of errors and in flow control. This is much less of a problem with long-distance optical fibre transmission. The delay for an optical fibre transatlantic link, for example, would be about 20 mS.

Note that the satellite communication is a **broadcast medium**. Many stations can transmit to the satellite and a satellite transmission can be received by many stations. This implies that digital satellite communication has much in common with local area network operation, which is also essentially a broadcast medium, and similar operating protocols can apply.

3.2 TRANSMISSION WAVEFORM

Representation of a digital signal for transmission can be made in a number of ways which we discuss later in section 4.2. The most obvious method is to transmit a square pulse to represent a binary 1, with a binary 0 represented by a zero level or a pulse waveform of opposite polarity (Fig. 3.5a and b). A particular significance can be attached to the amplitude of a pulse so that this can have one of N possible amplitude levels, thus permitting $\log_2 N$ bits of information to be conveyed by each pulse. Figure 3.5c illustrates this where four transmission levels are used (i.e. 2 bits of information per pulse).

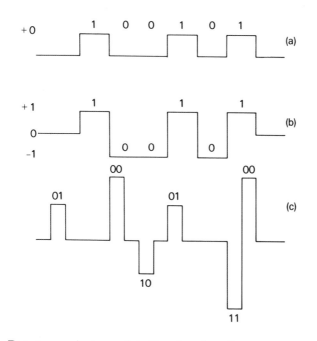

Fig. 3.5 Data transmission. (a) Two-level coding; (b) three-level coding; (c) variable-amplitude coding.

All these methods and others discussed later share a common characteristic, namely having a rectangular pulse waveform as the basic constituent for the transmitted information. If we know the way in which one such pulse passes through the transmission system then we can deduce its behaviour to a train of pulses constituting the data. A practical system will not deal very kindly with the transmission of digital impulse waveforms since it is not possible to achieve zero attenuation with distance, infinite bandwidth and no pulse distortion. It is necessary, therefore, to study the impairments to the transmission of a rectangular pulse that will inevitably occur in a practical system before we can see how this may be used to convey complete messages across the system.

3.3 TRANSMISSION IMPAIRMENT

A number of transmission limitations set a fundamental limit to the rate at which data can be transmitted through a medium and the accuracy with which it can be reconstituted at the receiving end. These are:

a attenuation,

b distortion,

c noise, and

d bandwidth.

For hardware media the signal attenuation is generally logarithmic and expressed as a constant number of decibels per unit distance. The attenuation of software media is proportional to the square of the distance from the transmitter, but for terrestrial transmission this can also be affected by atmospheric conditions. The installation of linear amplifiers (for analogue signals) or repeaters (for data transmission) at intervals along the transmission path operates to counter the effects of signal attenuation by restoring the signal to a level sufficiently above the accompanying noise level to permit the signal to be received without error. The total number of amplifiers and repeaters that can be employed in a given long-distance route is limited, however, by other forms of signal impairment, notably cumulative noise, and additionally for analogue signals, amplifier non-linearity. Consequently, there is a premium on the use of low attenuation media, such as fibre optics, which permit greater repeater spacing.

Attenuation in a given medium is also an increasing function of frequency, so that different spectral constituents of a signal behave differently with distance from the transmitter. Frequency distortion can be quite severe with analogue (and hence carrier) systems. Equalization techniques such as loading coils and filters are commonly introduced into line transmission systems to minimize these effects, but can lead to other deleterious results with digital transmission due to transient impulse signals initiated by the leading edge of the rectangular waveform[6]. A similar type of distortion, **delay distortion** also occurs with hardware media. This is caused by the variation with frequency of the velocity of propagation through the medium. We noted this in relation to fibre optics earlier. Delay distortion is particular serious with data transmission, where it causes the spectra associated with one digit pulse to affect the spectra of adjacent pulses. This is known as **inter-symbol**

interference and, as we shall see later, can be minimized by certain equalization techniques.

By far the most important of these transmission limitations are **noise** and **limited bandwidth**. We saw in section 2.3.4 how these are related to the transmission rate and how the choice of coding method can maximize the information rate for carrier systems. This also applies to direct data transmission, where certain coding methods can lead not only to optimum utilization of channel capacity, but also to methods of **error detection** and **error correction**. These will form the subject of the next chapter. Here the identification of different noise sources will be explained and some fundamental methods described that can be employed to match the data signal more closely to the characteristics of the transmission media.

3.3.1 Noise

The four generally recognized categories of noise in data transmission are[7]:

a thermal noise,

b Gaussian noise,

c jitter noise, and

d impulse noise.

Carrier systems are also subject to **intermodulation noise** which occurs when signals of different frequencies share the same transmission medium and encounter some non-linearity in the system, e.g. a non-linear amplifier. In addition, adjacent metallic conductors can be affected by **cross-talk** due to electrical coupling existing between neighbouring transmission lines.

Thermal noise is due to thermal agitation of electrons in a conductor. It has a uniformly wide spectrum and is a function of temperature. Although it cannot be eliminated and represents an upper bound on communications system performance, its actual level is very small (about $-100\,dB$ for a bandwidth in the tens of megahertz region) and can be ignored for most applications.

Additive **Gaussian noise** occurs in any transmission system incorporating amplifiers. In the PSTN such noise can be heard as a background 'hiss'. This noise usually has a zero d.c. value and has an approximately Gaussian amplitude distribution. For a given noise threshold level there is some particular probability that the noise will add to the value of the signal and turn a correct signal into an incorrect one. Noise will thus cause a proportion of the symbols received to be in error. Gaussian noise is very dependent on this threshold and beyond it the error rate increases considerably. For a leased line in the PSTN it is at about $-45\,dB$ relative to $1\,mW^2$.

Much more serious for data transmission is **impulse noise**. This consists of irregular pulses or pulse 'bursts' such as radiated ignition interference. The interfering pulses are often of high amplitude, and although of limited duration they can encompass very many consecutive bits of data and thus irreparably damage a transmission frame[1]. In such a situation the damaged frame would be detected and a request forwarded for retransmission. Methods of **flow control** have been developed to check the data received and to initiate the automatic retransmission of the defective information. This method is clearly inefficient when long frames of data or lengthy message transmission are attempted and is one of the factors influencing the decision to transmit the data as a series of short **packets** of information instead; a whole set of such packets constituting the message. Forms of signal encoding may then be used which facilitate automatic error detection. These will be considered later in section 4.4.

3.3.2 Signal-to-noise ratio

The parameter used for determining the noise performance of a transmission is **signal-to-noise ratio** (SNR). If we denote the signal power by S and the noise power by N, the signal-to-noise ratio is S/N. Usually the ratio itself is not quoted, instead the quantity is expressed in decibels as:

$$SNR = 10 \log_{10} \frac{\text{signal power}}{\text{noise power}} \text{ dB}$$

(3.5)

and gives the amount in decibels by which the signal(s) exceed the noise level in the transmission environment. The SNR is especially important in the transmission of digital data since it determines the maximum data rate possible in a given situation (see eqn 2.7).

3.3.3 Phase jitter

A continuous series of short-term departures of the individual transmitted pulses from their correct time position is known as **phase jitter**. When isolated from the affected pulses, phase jitter shares many of the characteristics of electrical noise and can be regarded as a phase-shift noise additive to the signal. Figure 3.6 illustrates the effect of jitter on a pulse train for a deterministic sinusoidal variation. In practice, the phase noise is much more likely to be non-deterministic and irregular.

A major source of jitter lies in the process of digital regeneration within repeaters spaced at intervals along a transmission medium. This kind of jitter is dependent on the pattern content of the transmitted signal and is often termed **pattern-dependent jitter**[8]. It is important because the effect of such a

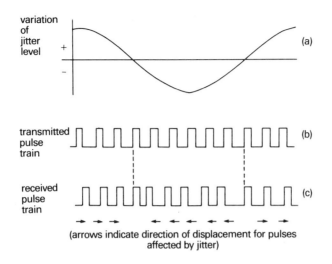

(a)

(b)

(c)

(arrows indicate direction of displacement for pulses
affected by jitter)

Fig. 3.6 Jitter affecting a train of impulses.

jitter is increased or accumulated with each passage through a repeater. It has been found to increase its value approximately proportional to the square root of the number of regenerative repeaters. Other forms of jitter are completely random in form and related to cross-talk, multiplexing, temperature variations and other effects.

Two basic methods of minimizing jitter are in use. The first involves the use of a digital scrambler mentioned earlier which, by introducing a random element in the signal, reduces the pattern-dependent action of the regenerator. A second is a jitter-reducing circuit which is effectively a re-timing circuit having a bandwidth that is small compared with the jitter bandwidth of the signal. Both methods are found in digital communications equipment, particularly that associated with high-speed transmission such as computer and satellite communication systems[9].

3.4 INTER-SYMBOL INTERFERENCE

Ideally, pulse transmission demands an infinite bandwidth to permit perfect pulse reproduction at the end of a transmission link. A major effect of a practical **limited bandwidth** communication system is the existence of a differential delay distortion for different components of the pulse spectrum arriving at the receiving end.

Let us consider first a channel having ideal characteristics except that frequencies beyond some frequency limit, B, are completely attenuated; thus it will not transmit a short pulse without *some* distortion. The response to such a channel has the form shown in Fig. 3.7a and is expressed mathematically as an **impulse response function**, $h(t)$:

$$h(t) = \frac{\sin(2\pi Bt)}{2\pi Bt}$$

(3.6)

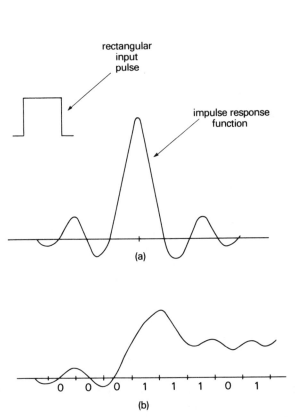

(a)

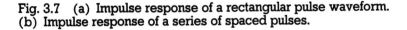

(b)

Fig. 3.7 **(a)** Impulse response of a rectangular pulse waveform. **(b)** Impulse response of a series of spaced pulses.

Thus if a short rectangular pulse is applied to the input of an ideal band-limited channel, the waveform at the channel output will have the form of a rounded pulse waveform with oscillating tails. The problem arises when we transmit a succession of pulses, since the 'tails' caused by previous pulses may obscure the main response of the current pulse. This effect is known as **inter-symbol interference** and is illustrated in Fig. 3.7b where a 0 following a string of 1s would be interpreted as an unbroken string of 1 values with the 0 value lost in the overall reproduced waveform.

We can, however, take advantage of the regularity of these oscillations to avoid inter-symbol interference altogether. We do this by sending pulses with a pulse spacing of $T = 1/2 B$. That is, if the cut-off frequency of the channel is B Hz, we send $2B$ pulses per channel. This is because the channel impulse response, $h(t)$, is zero at the sampling instants $t = +1/2 B, +2/2 B, +3/2 B$, etc. as shown in Fig. 3.8, indicating that at $t = n T$, the sampled channel output is due solely to the main lobe of the response of the channel to the nth input pulse; the response of the channel to all the other input pulses is zero at this instant.

three adjacent sampling points

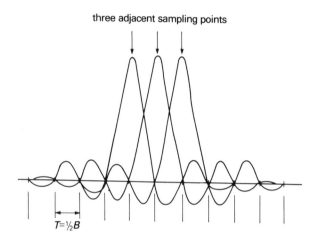

$T = \frac{1}{2} B$

Fig. 3.8   Critical spacing applied to minimize inter-symbol interference.

In practice it is difficult to construct band-limiting filters giving a good approximation to the $(\sin x)/x$ impulse response and, in any case, a system such as this would be extremely sensitive to small timing errors. What is needed is a channel whose impulse response has the same property as the $(\sin x)/x$ function, i.e. providing zero value at $t = +T, +2T, 3T$ etc., but with a gentle roll-off with frequency and capable of being realized with practical filter design.

Nyquist[10] has shown that if the channel frequency characteristic, instead of cutting off sharply at some limiting frequency, B, exhibits a symmetry about a frequency equal to half the pulse transmission rate, and also has a linear phase characteristic, then the impulse response of the channel will exhibit a succession of zeros at the appropriate points.

A suitable response which can closely approximate this in practice is the **raised cosine roll-off** function shown in Fig. 3.9. Where $K = 0$ then this is the familiar $(\sin x)/x$ response. Where $K = 1$ then not only are the tail perturbations reduced but an additional zero point is included midway between sample values. The price paid for this is an increased bandwidth (for $K = 1$ this is doubled). In a practical application of these concepts an **equalization circuit** is interposed between the data signal to be transmitted and the transmission system. This consists of a **transversal filter**[11] designed to modify the rectangular pulse waveform so that the edges of the waveform conform to the raised cosine shape and so match the frequency characteristics of the channel. Since these may not be known precisely at the time of transmission, an **adaptive equalization** process is often used. Here, a test pattern is sent prior to the data and the *a priori* knowledge of this pattern at the receiver is used to compute the impulse response from the received signal. It is then possible to adjust the filter characteristics directly or to use an iterative technique to successively increment these until an optimum setting is realized[12].

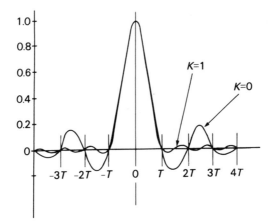

Fig. 3.9 A raised cosine roll-off function.

3.4.1 The eye diagram

Inter-symbol interference can be readily examined by arranging the synchronous superposition of all possible data patterns along a time axis. The waveform that leaves the equalizing filter can be displayed in this way and will include all the sources of error. Because of the different states of each bit and its neighbouring bits there will be many possible signal waveforms. When these are superimposed through synchronization of the waveform with the clock interval and displayed by using an oscilloscope, a waveform such as Fig. 3.10a is obtained. This shows an accurately balanced waveform with the 'eye' of the diagram (the shaded area) open. The presence of severe inter-symbol interference will cause the eye to close as shown in Fig. 3.10b, due to the random displacement of successive bit response waveforms. The eye diagram thus provides an on-line visual check in the time domain of the effective channel frequency response.

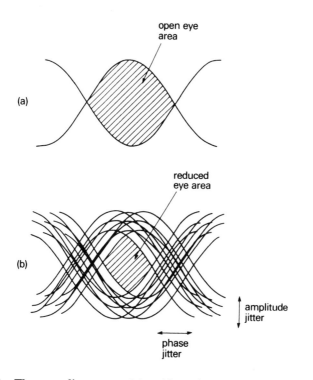

Fig. 3.10 The eye diagram: (a) with no inter-symbol interference;
(b) with considerable inter-symbol interference.

SUMMARY
Chapter 3

The characteristics and limitations of various forms of transmission media are described for analogue and digital transmission. An emphasis is placed on transmission using fibre optic cables due to its high bandwidth capability and low attenuation.

A number of signal impairments occur during transmission, principally noise and attenuation. These can be minimized by choice of transmission media, coding methods and the use of repeaters at intervals along the transmission path.

Gaussian noise and impulse noise are particularly damaging to the transmission of digital data, so that methods of error detection and correction are required in a practical data communication system. Transmission of essentially infinite bandwidth digital pulses over a medium having a limited bandwidth can cause a particular form of interference, known as inter-symbol interference, to occur between consecutive pulses in the transmitted message. The effect of this can be reduced by modifying the characteristics of the digital pulses before transmission using various forms of equalization circuit.

PROBLEMS
Chapter 3

P3.1 **a** What are the advantages of using optical fibres in a communications system? Describe some practical difficulties in implementing a fibre optic network covering an area of several hundreds of square kilometres.

b Referring to Fig. 3.2, prove the relationship between angle Θ of the propagating light ray to the core axis and the angle $\varnothing$ at the core-cladding interface.

P3.2 An optical fibre is designed to operate as a multimode transmission device. The core refractive index is 1.54 and the outer cladding refractive index is 1.52. It is found to have a loss of 4 dB for each kilometre used. If the maximum data transmission rate is 2 Mbps determine:

a The maximum fibre length possible for minimum pulse spreading.

b The decibel loss for this length.

(The velocity of light may be taken as 3×10^8 m/s.)

P3.3 A graded optical fibre cable has a maximum refractive index of 1.5 with $d = 0.1$.

a Compare the pulse spreading obtained for a 10 km cable with a similar length of multimode optical fibre cable having a refractive index ratio of $n_1/n_2 = 1.05$ and an inner core diameter of 50 mc.

b What do you deduce concerning the performance characteristics of these two cables from the figures obtained for pulse spreading?

P3.4 Discuss the choice of transmission frequencies for a space satellite communications system. How is bidirectional communication achieved? Contrast the communications problems for speech transmission via a satellite with local speech communications over the telephone network.

P3.5 What is signal-to-noise ratio? In a transmission system several sources of Gaussian noise are identified and measured in terms of their power level at different points in the network. These are:-

(1) 10 mW at the output of the transmission equalizer

(2) 5 mW contributed at each of the 20 repeaters between the transmitter and the receiver

(3) 10 mW contributed by the receiving equipment

(4) At certain times of the day a further 30 mW of noise is also induced in the line.

If an operating signal level of 500mW is essential at the output of the receiver, what is the minimum signal-to-noise ratio that can be tolerated.
a generally and
b at peak interference times?

P3.6 A transmission system has a measured signal-to-noise ratio of 25 dB. What is the minimum bandwidth needed if a channel capacity of 50 kbps is to be realized with an added deterioration of 10 dB due to the phase jitter known to be present with the form of modulation used. (Hint: use eqn 2.7 in your answer.)

P3.7 State the difference between an amplifier and a repeater used in digital communication networks. Whey are they necessary? What limits the number that may be used in a long-distance network?

4

CODING AND FRAMING

Given a suitable transmission medium we next consider how to use this effectively for conveying the data, consistent with the need for accuracy and with regard for the quantity of data to be transferred. In Chapter 2 the use of an analogue carrier system was considered, involved the use of modulation to **encode** the digital **source data** onto a continuous analogue signal, the **carrier**, using **frequency division multiplexing** to transmit many channels of data over a single communications link.

Carrier systems involving a continuous analogue signal are gradually being replaced by digital methods in which **time division multiplexing** is used to provide the high-density transmission needed. The device used to convert analogue signals, e.g. telephone conversions, into digital data for transmission over such a system is known as a **codec** (coder/decoder), which plays a similar role to the **modem** in an analogue carrier system. The principal technique used in the codec is **pulse code modulation** (PCM) which will be considered briefly in this chapter.

In direct digital transmission, as well as in PCM, binary data are transmitted by encoding each data bit into signal elements. The encoding scheme used for this process is known as **source coding** and represents a mapping from individual data bits to signal elements. The way in which this is carried out can affect substantially the ability of the system to interpret the signal and to recover from errors incurred in transit. A number of the more common schemes in use are described below.

As discussed in Chapter 3, there are many possible impairments preventing accurate reception of the transmitted data. Methods of identifying (trapping) these errors have been developed which are widely used in data transmission. Fundamental to the **error detection process** is the subdivision of the stream of transmitted data bits making up the message into smaller groups or frames of digits. In section 2.6 this was considered for channel sequence synchronization. Here the framing concept is applied at the more fundamental bit level. Transmission of frames of data thus serves to delineate sets of digital data and provides a structure for synchronizing the reception process with the expected duration and time of each bit transmitted.

Finally, the basic mechanisms for data transfer of this encoded information over a transmission medium are considered. Whenever two digital devices, e.g. a terminal and a computer, are linked over a transmission network, a considerable degree of cooperation is required. There needs to be some agreed **control** arrangement to ensure that the timing, rate and duration of the transmitted bits is understood correctly and an **identification** scheme agreed whereby the function of the various bits constituting the frame may be recognized. To facilitate this control the digital data device is not connected to the transmission medium directly but through a standardized **interface**. Several of the common standards used for interfacing are considered.

4.1 PULSE CODE MODULATION

Signal representation for analogue signals transported over a digital channel requires a conversion process which involves three elements:

a   sampling in the time domain,

b   quantization in the amplitude domain, and

c   coding into digital form.

All three processes impose limitations in their application and can give rise to various sources of error.

4.1.1 Sampling

Sampling involves the selection of a series of narrow impulses or 'slices' of the signal, usually spaced at equal time intervals. This was illustrated earlier in Fig.1.5 as part of the pulse code modulation process. A unique number is ascribed to each impulse and represents the mean amplitude of the sample taken over the area of the impulse. The slice should be infinitely narrow but, in a practical case, it is necessary to estimate an averaged quantity over the sampling period. The length of time over which the data are averaged is known as the **aperture**. Aperture errors caused by the amplitude value varying over the sampling interval are minimized by incorporating very fast multiplexing and analogue-to-digital converters within the codec device. Additionally a technique, referred to as **sample-and-hold**, may be employed where the signal valve at the leading edge of the sample is maintained over the entire sampling period.

4.1.2 The sampling theorem

Some care has to be taken in choosing the **rate** of sampling for the analogue signal in order to avoid sampling errors. The difficulty is illustrated in Fig. 4.1. Here, the same finite set of sampled values, taken at regular intervals, are seen to fit a number of arbitrary signals which, after quantizing and coding, are indistinguishable from each other. (Remember that the process of sampling can only be approximation since the amplitude values *between* the sampling pulses are unrepresented.) We say that these are **aliased** signals. Thus, for a signal, $x(t)$, representing a waveform, $\cos 2\pi f_0 t$, then a set of aliased signals having different frequencies can be shown to exist which are related to the sampling interval, h as:

$$f_0, (1/h)-f_0, (1/h)+f_0, (2/h)-f_0, (2/h)+f_0 \ldots (n/h)\pm f_0 \qquad (4.1)$$

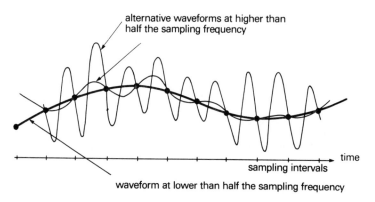

alternative waveforms at higher than half the sampling frequency

time

sampling intervals

waveform at lower than half the sampling frequency

 •=sampling points

Fig. 4.1 Aliasing.

The range of frequencies below which this effect is not present extends from $f_0=0$ to $f_0 = f_n$. This maximum frequency, f_n is the **Nyquist frequency** and is a frequency limit (the **Shannon limit**) to the sampled data, above which an unambiguous reconstruction of the signal from the sampled values is not possible.

Hence, given a signal having a bandwidth, B Hz, and containing no frequency components at and beyond a frequency f_n, then the lowest sampling frequency necessary to preserve the information contained in a sampled version of this signal is given as $f_s > 2B$ or, since $f_s = 1/h$, then the minimum bandwidth necessary $B = 1/2h$. This is known as the **sampling theorem** and directly relates the sampling interval to the signal bandwidth.

It follows that for a given frequency spectrum, the individual frequency components lying between $f=0$ and $f=B$ can be separately examined but, if the signal contains components having frequencies $f>B$, they will not be distinguishable and will simply be added to the lower frequency signals and contribute to the noise level associated with the signal. To summarize, the sampling theorem consists of two statements viz:

a Signals having a finite bandwidth up to and including B Hz can be completely described by specifying the values of the signal at particular instants of time separated by $1/2B$ s.

b If the signal is band-limited and contains no frequency greater than B Hz, it is theoretically possible to recover completely the original signal from a sampled version when the sampling interval is equal to or smaller than $1/2B$ s.

To ensure that aliasing error is not present with the sampled signal, we need to limit the bandwidth of the signal to be below B and this is carried out through a low-pass filtering operation *before* digitization, and sampling at a frequency at least twice the highest frequency of interest in the signal.

This sampling process is referred to as **pulse amplitude modulation** (PAM) and is the first step towards the PCM process. To produce PCM data the PAM samples must first be quantized.

4.1.3 Quantization

Representation of a variable-amplitude series of discrete sampled values as a limited series of discrete numbers is termed **quantization**. The process can only be an approximation since, whilst the original signal can assume an infinite number of states, the number of bits in a digital representation is limited. The resultant numerical value is given as an integer corresponding to the nearest whole number of units. This follows from the transfer characteristic of such a quantizer which is shown in Fig. 4.2. An input value lying between the midpoint values of two consecutive unit values will produce an output at a level corresponding to the higher of the two values. The uncertainty or distortion introduced by this method of sample level determination gives rise to a **quantization error**, which is defined as the difference between the discrete value of the signal at a sampling instant and its nearest quantized value. It is usually expressed as a form of noise added to the signal and referred to as **quantization noise**[1]. This is related to the number of quantization steps, N, taken to represent the dynamic range of the signal being quantized and is given approximately as[2]:

$$SNR_q = 6\log_2 N + 1.8 \quad dB \tag{4.2}$$

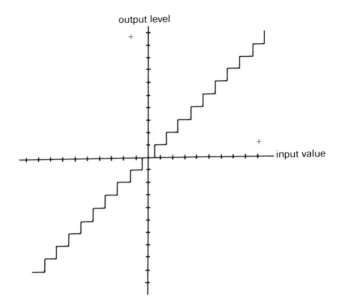

Fig. 4.2 Quantization transfer characteristics.

where SNR_q is a signal-to-quantization noise ratio. Note that this ratio improves by 6 dB for each extra bit allocated to the uniform quantization process. Thus, if the signal is to be quantized to at least $N=256$ possible levels (8 bits) then the SNR_q achieved will be approximately $6 \times 8 + 1.8 = 49.8$ dB. This level of noise is found acceptable for speech transmission.

If we take this quantization level of 8 bits as an adequate representation of a speech signal (0 – 4 kHz), and carry out signal sampling in accordance with the sampling theorem at $2f_{max}=8$ kHz, then the data transmission rate will be 64 kbps. This, as we shall see later, has been accepted as an important CCITT recommendation, I-420, applied to the **Integrated Services Digital Network** (ISDN) for a combined speech/data transmission system.

4.1.4 Non-linear coding

In PCM transmission of speech there are advantages in sampling not at regular intervals, as shown in Fig. 1.5, but at non-uniform intervals. This is to allow small-amplitude signals to be more finely quantized and so improve the quality of the coded speech without increasing the number of bits per sample. Three methods of non-uniform quantization are in use. They are:

a The decision values are linear, but the analogue signal is **compressed** at the coder input and **expanded** at the decoder output. This was the original method but is now no longer used since it is less easy to implement with the same degree of precision as later methods (**b**) and (**c**).

b The decision values are non-linear. In this method the intervals between the decision values are made smaller for low-level signals, than those for high-level signals.

c The decision values are linear, but with the smallest necessary. quantum step used over the whole dynamic range of the encoder. The required non-linear characteristic is then achieved by manipulation of the binary numbers, reducing the total number of digits to the same as that used for method (**a**).

A widely accepted encoding law is the CCITT recommendation G732 for a 30-channel system[3]. This approximates to a continuous function given by:

$$y = \frac{1 + \log_e(Ax)}{1 + \log_e A} \quad \text{for } \frac{1}{A} < x < 1 \tag{4.3}$$

$$y = \frac{Ax}{1 + \log_e A} \quad \text{for } 0 < x \frac{1}{A} \tag{4.4}$$

where A = 87.6 and is shown in Fig. 4.3.

A segmented approximation to this continuous function can be drawn, so that each successive segment changes its slope by a factor of 2 which we see from the dotted line in the diagram.

In this example the positive input of the applied signal, $x(t)$, is allocated 128 decision values for the quantized output samples Y_i, thus requiring seven binary digits. The negative input range is similarly quantized. In practice, an eight-digit code is applied where one binary digit is used to indicate the polarity of the sample and the remaining seven digits are used to indicate the quantized ouput values for both positive and negative samples.

The effective reduction in the number of bits required for a non-uniform quantizer compared with a uniform quantizer having the same dynamic range, is known as the **companding advantage** and is given as:

$$20 \log_{10} \frac{N}{n} \text{ dB} \tag{4.5}$$

where N = number of decision values of the uniform quantizer and n = number of decision values of the non-uniform quantiser.

For the A-law characteristic shown in Fig. 4.3 this is approximately 24 dB.

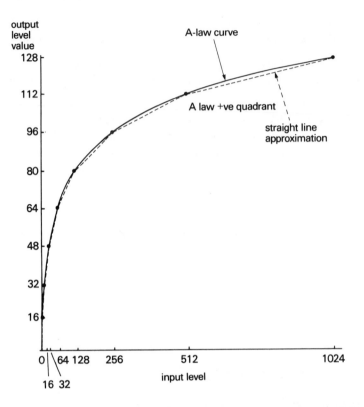

Fig. 4.3 *A*-law encoding characteristic for the CCITT 30-channel system.

4.2 SOURCE CODING

The digital signal to be encoded for transmission consists of a sequence of discrete discontinuous pulses or bits of data. When these are encoded they are referred to as **signal elements**. In the simplest case, there is a one-to-one correspondence between bits and signal elements, in which, for example, a binary 0 is represented by a lower voltage level and a binary 1 by a higher voltage level. A variety of other **encoding schemes** are in use which are discussed in this section. Some of the more common are:

a non-return to zero (NRZ),

b return to zero (RZ),

c bi-polar, and

d bi-phase (Manchester).

4.2.1 NRZ coding

NRZ coding (Fig. 4.4a) maintains a constant voltage level during a bit interval. This is the simplest code to generate and makes efficient use of the available bandwidth. It does provide some difficulties in use, however. The encoded waveform has a significant d.c. component which is a problem with transformer connection. Further, NRZ has few zero transitions from which a timing clock signal could be derived. A long string of 1s would thus be difficult to detect in the presence of voltage drift at the receiver or transmitter. Two variants of the NRZ code are in use. These are NRZ-M in which the level is changed only when a transition from mark to space occurs, and NRZ-S for a transition from space to mark. These result in a level transition for each bit and effectively remove the d.c. component. However, the bandwidth is considerably increased with these methods, making the coding less efficient.

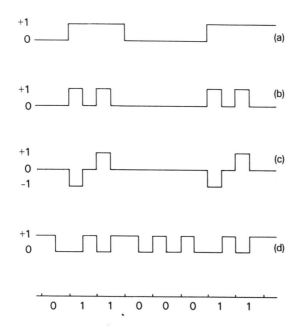

Fig. 4.4 Source encoding schemes. (a) NRZ coding; (b) RZ coding; (c) Bi-polar coding; (d) Bi-phase (Manchester) coding.

4.2.2 RZ coding

As shown in Fig. 4.4b, RZ coding provides a return-to-zero level in the middle of the pulse period. This results in a symbol rate (in bauds) of $2/T$ which thus differs from the bit rate $1/T$ (in bps), where T is the bit duration.

Although RZ enables a continuous series of 1s to be more readily detected, it still does not solve the problem with the transmission of a string of 0s, which are still transmitted as a single level. It also results in an increase in transmission bandwidth for very little advantage.

4.2.3 Bi-polar coding

The third alternative shown in Fig. 4.4c is known as a **bipolar code** and successfully overcomes these difficulties. Here the polarity of alternate pulses representing digital 1s are inverted, irrespective of the number of zeros between them. This code is also referred to as a **alternate-mark inversion** code. The average value of the signal is now zero and hence there is no d.c. component. These codes maintain a constant voltage level during bit interval, i.e. there is no return-to-zero level during the transmission of a bit. Because it is easy and unambiguous to detect these transmitted pulses this code is widely used by low-speed data processing terminals and similar devices.

A significant disadvantage of bi-polar coding is that long sequences of zeros will inhibit the working of timing recovery circuits. A version of the bi-polar code, known as **high-density bi-polar code** and designated as **HDB3 code** has been devised to overcome this[4]. Here the maximum number of consecutive zeros that can occur is limited to three. Sets of four consecutive zeros are replaced by 000V where V represents a violation of the bi-polar coding, i.e. a digital 1 is deliberately inserted which has the *same* polarity as the last digital 1 transmitted. (As shown in Fig. 4.4c, we would expect a digitial 1 in the coding always to have the opposite polarity from the previous digital 1.) Since bi-polar detection circuits are arranged to detect only *alternate* polarity pulses, this violation pulse can easily be recognized so that after extraction of the clock pulse the bit stream 000V can be replaced by the four zeros again.

4.2.4 Bi-phase (Manchester) coding

An alternative to bi-polar coding is **bi-phase encoding** which only requires two coding levels and is thus easier to implement than bi-polar coding, which is essentially a ternary code. Several versions of this exist of which the most widely used is called **Manchester encoding**. This requires two transitions for each bit and hence requires an increased bandwidth over NRZ coding. It does, however, overcome the problems of NRZ and RZ coding without the complexity of high-density bi-polar decoding logic. As shown in Fig. 4.4d, the time interval available for each digit is divided into two halves with a level transition *always* occurring in the centre. Thus a 1 bit is transmitted as a zero level, followed by a one level, and a 0 bit as a one level followed by a zero level. A considerable advantage of this scheme is that because there is a predictable transition during each bit time, the receiver can synchronize on that transition. For this reason the bi-phase codes are known as **self-clocking codes**. Manchester encoding is widely used for computer communications, particularly in local area networks, due to its ease of implementation and cheapness.

4.3 SYNCHRONOUS AND ASYNCHRONOUS TRANSMISSION

Of fundamental importance in any scheme of bit transmission using single-channel or multiplexed transmission is to devise a method to enable the receiving end to know the starting time and duration of each bit transmitted. Without this, the information transmitted would become a meaningless jumble when attempts are made to decode it at the other end of the transmission link.

The key to these synchronizing methods is to define the data in terms of small blocks, i.e. **frames**, and to delineate the start and stop of the frame so as to permit resynchronization at the beginning of each block. The simplest scheme is actually referred to as **asynchronous transmission** and is shown in Fig. 4.5a. Data are transmitted one character (usually 8 bits) at a time. Each character is preceded by a **start code** and followed by a **stop code**. These can be a single 0 and 1 digit, respectively, and when no data are sent the transmitter sends out a continuous stop code (i.e. a series of 1s). The method is termed asynchronous due to the variable and unpredictable gap that exists between the data characters transmitted.

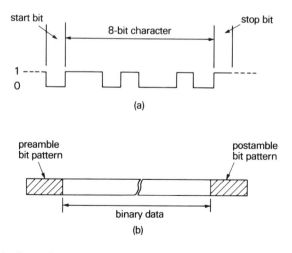

Fig. 4.5 (a) Asynchronous transmission. (b) Synchronous transmission.

For computer communications this mode of operation is unnecessarily wasteful. A more efficient method is **synchronous transmission** (Fig. 4.5b). Here a continuous binary data stream is transmitted without start or stop codes but with the exact transmitting or arrival time of each bit known. This implies that the transmitter and receiver clocks must be somehow synchronized and means made available to determine the beginning and end of a data message. To achieve this each block begins with a **preamble** bit pattern and ends with a **postamble** bit pattern. These represent unique and significant bit patterns which can be recognized and acted upon[5]. Since synchronized transmission does not distinguish between bits or characters contained in a block, it is often called a **bit-oriented** scheme with asynchronous operation as a **character-oriented** scheme.

4.3.1   Bit stuffing

One problem that can arise with bit-oriented transmission is that the significant bit patterns used for sync recognition can also occur somewhere else in the data. In one widely used scheme for synchronous data transmission a bit pattern 01111110 is used for both preamble and postamble bit pattern. To avoid the appearance of this pattern in a data stream, the transmitter automatically inserts an extra 0 bit after each occurrence of five 1s in the data being transmitted. When the receiver detects a sequence of five 1s it examines the next bit. If this is 0 the receiver deletes it. If it is a 1 then the bit pattern must be part of the preamble code. This procedure is known as **bit stuffing** and has some similarities with the procedure of **pulse stuffing** which we met earlier in connection with asynchronous TDM (section 2.6.4).

Synchronous transmission is far more efficient than asynchronous working when applied to sizable blocks of data. This is particularly the case for 'bursty' traffic which occurs with computer communications, when the additional overhead required of asynchronous operation could reach 20% or more. The control information in synchronous transmission is typically less than 100 bits, which is acceptable for messages greater than about 1000 bits. For example, in one of the more common bit-oriented schemes, HDLC, which we consider later, 48 bits of control information are included in each frame. Thus for a 1500 bit message, the overhead is only $48/1548 \times 100\% = 3.1\%$.

4.3.2   Clock extraction

With **asynchronous transmission** a separate clock is used whose frequency is typically several times higher than the transmitted bit rate. It is then possible to detect the leading edge of the start bit at the front of each character bit set and use this, in conjunction with the receiver local clock, to estimate the centre of each bit period. Since this occurs with each character, a cumulative phase shift or error is unlikely to occur. We cannot do this with **synchronous transmission** since start and stop bits are not used and the *whole* of the frame is transmitted continuously as a stream of bits.

Two methods of clock extraction are used. The clock information can form part of the transmitted bit stream and is extracted by the receiver. This is applied in line coding through the bi-polar HDB3 coding scheme discussed earlier. Secondly, the encoding method for the information can be arranged such that there are enough guaranteed transitions in the bit stream to synchronize a separate clock at the receiver. This is the case for bi-phase Manchester encoding.

In operation, the timing signal recovered from the receiver data bit stream will have a rate corresponding to the average rate of the line signal. The signal is operated upon by a full-wave rectification and slicing circuit to produce a spectral component at a frequency equivalent to the bit rate. In turn this is applied to a resonant circuit tuned to this frequency, and after amplification and differentiation accurate timing pulses can be generated to re-time the clock signal. The resonant properties in the recovery tuned circuit will enable correct synchronization to be maintained in the absence of a short period without timing transitions in the bit stream so that it may be applied very successfully to bi-polar and HDB3 coding[6].

4.4 ERROR CHECKING AND CONTROL

Transmission errors resulting in missing or altered digits are inevitable in any realizable system. To overcome this we need to apply error detection checks and, if possible, to initiate error correction procedures. Error detection usually means adding one or more bits to each frame transmitted. These additional bits constitute an **error detecting code** and the principal codes used are **parity checks** and **cyclic redundancy checks**.

Error correction is more difficult and not often used in data transmission since it is easier to retransmit a faulty frame. Such error correcting schemes that are in use often represent a reduction in effective data rate of 50% and are only applied in those situations where retransmission is impossible (e.g. some space missions)[7].

4.4.1 Parity checking

The simplest bit error detection scheme is to append an extra bit to the frame in such a way as to produce a predictable number of even or odd number of 1s in any frame. This extra bit is called a **parity bit**. A typical example is ASCII character transmission in which a parity bit is attached to each 7-bit ASCII character to make an 8-bit word. The value of this bit is selected so that the word has an even number of 1s (even parity) or an odd number of 1s (odd parity). Typically, even parity is used for asynchronous transmission and odd parity for synchronous transmission.

For example, if the transmitter is transmitting an ASCII G character (1110001) and using odd parity, it will append a 1 to this and transmit 11100011. The receiver examines the received character and, if the total number of 1s is odd, assumes that no error has occurred. If one bit or any odd number of bits is erroneously changed during transmission (e.g. 11000011), then clearly, the receiver will detect an error. Note that if two (or any even number of) bits are changed, the error goes undetected. Apart from the problem of an even number of changed bits, parity checking can be badly affected by additive noise having a period lasting longer than that of a single bit.

An improvement may be obtained by using **block parity checking** shown in Fig. 4.6. This involves a second set of parity bits acting along the *columns* of a block of data. Using both **longitudinal parity bits** and **vertical parity bits** the error rate can be reduced by about 2–4 orders of magnitude compared with simple longitudinal checking[8].

Information bits

0		0	0	0	0	0	0	0	
0		0	0	1	1	0	0	0	
1		0	1	0	0	0	1	1	
0		1	0	0	1	0	0	0	Even Parity
1		0	0	0	0	0	1	0	
0		1	0	0	0	1	1	1	
1		1	1	0	0	0	1	0	
1		1	0	1	0	1	0	0	

Row Check bits

Column Check bits

Fig. 4.6   Block parity checking.

4.4.2   Cyclic redundancy checks

A more complex procedure which finds wide use in computer communication is **cyclic redundancy checking** (CRC). Given a k-bit **message sequence**, the transmitter generates an n-bit sequence, known as a **frame check sequence** (FCS), r, so that the resulting frame, consisting of $k+n$ bits is exactly divisible by some predetermined bit **division sequence**, p. The receiver then divides the incoming frame by p and, if there is no remainder, assumes that no error is present.

How this works is best seen on an actual example in which we use **modulo-2 arithmetic**. Modulo-2 arithmetic uses binary addition with no carries and is effectively an **exclusive-OR** operation in logical terms. For example, modulo-2 addition and multiplication may be shown in:

$$
\begin{array}{r}
1111 \\
+1010 \\
\hline
0101
\end{array}
\qquad \text{and} \qquad
\begin{array}{r}
11001 \\
\times \quad 11 \\
\hline
11001 \\
11001 \\
\hline
101011
\end{array}
$$

As we stated above, in order to achieve the correct check result we need to make $(k+n)/p$ equal to a whole number (i.e. containing no remainder after division). What actually happens is that a suitable value of p is chosen, one bit longer than r (the frame check sequence we need to calculate). The message is divided into p to obtain a **remainder**, which is the frame check sequence r. This is appended to the transmitted message to give a frame sequence:

Transmitted frame $= k + r$ (4.6)

The division by p (the value of which is made known to the receiving station) is then carried out upon receipt of the frame, and if a zero remainder is obtained, an error-free transmission is assumed.

As an example, if we let the message $= k = 11011$ and the divisor $= p = 11001$ (five bits), then the remainder, r, will be 4 bits long. The steps are:

1 k is multiplied by 2^4, i.e. shifted four places to the left to make room for the FCS to be added later to give $k' = 1100110000$.

2 k' is divided by p, i.e.

$$
\begin{array}{r}
100001 \\
11001 \overline{)\ 1100110000} \\
11001 \\
\hline
\end{array}
$$

(modulo-2 subtraction)
$$
\begin{array}{r}
10000 \\
11001 \\
\hline
\end{array}
$$

10001 = remainder = r

3 This is added to the augmented message, k' to give $k' + r = 1100111001$. This is the transmitted frame.

At the receiving end the frame is divided by p to yield:

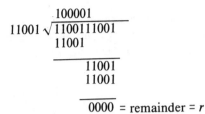

Since r is found to be zero, then no errors are assumed to have occurred.

Should the received message become corrupted during transmission, for example through an error burst making the three least significant digits all 1s so that the received frame is equal to 1100111111, then the division by p yields:

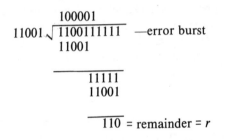

r is now no longer zero and an error is detected.

The divisor, p, sometimes referred to as an **encoding sequence**, depends on the type of errors expected. For example, an error pattern that is identical to the divisor will generate the same check bits as in correct transmission and hence be undetectable. A divisor which is prime, in the modulo-2 sense, is therefore normally chosen.

4.5 INTERFACING

Equipment between which data transfer is to take place is referred to as **data terminating equipment** (DTE). This includes not only computers but also a wide range of devices such as visual display units, work stations, printers, plotters etc. The DTEs are connected together not directly but via an **interface** to carry out the actual data exchange. This is effected through a **communicating protocol**, which is essentially a set of conventions or rules agreed by both parties in the exchange to ensure that data being exchanged is interpreted correctly. Thus, to control the order and timing of data transfer between DTEs a set of control signals becomes necessary. At its simplest level, being concerned only with getting a bit stream from one device to another, these control signals are said to form the **physical level protocol**, which, we will see later, is the lowest level in the more complex seven-level OSI protocol model which forms the subject of Chapter 8. Two of these level 1 interfaces will be described, one for telephone carrier transmission, i.e. modem control, known as RS-232c and another for digital data transmission called X21.

4.5.1 RS-232c interface

This interface is defined by the **Electronic Industries Association** in the USA as RS-232c and, alternatively, as the V24 interface by the **CCITT** in Europe. It was originally defined as the standard interface for connecting a DTE to an approved modem. It is now used generally to connect to any character-oriented peripheral such as a printer, visual display unit, computer or another terminal.

Where a connection is made to a transmission line, the interface is connected between the DTE and the control equipment to which the line is connected. This latter is known as the **data circuit terminating equipment** (DCE). These connection arrangements are shown in Fig. 4.7. In order to arrange the interchange of information for a variety of devices, some 37 different coded electrical control signals are defined where logical $1 = < -3\,V$ and logical $0 = > +3\,V$. In general, the signals convey information in one direction between the DTE and the DCE. We can group these control signals conveniently into three groups: data signals, control signals, and timing signals. These are shown in Table 4.1. A full list and description is given in ref.[9].

The functions of most of this limited subset are fairly obvious from their titles. The ring indicator (CE) has a parallel with telephone practice and alerts the DTE that a calling signal is being received on the communication channel.

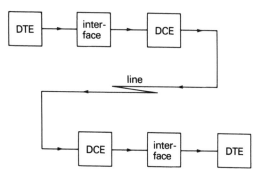

Fig. 4.7   DTE–DCE connection.

Table 4.1

RS-232c interface signals

	RS.-232c	V24	Pin No	Direction DTE	DCE
Group A: Data Signals					
Transmitted data	BA	103	2	-------->	
Received data	BB	104	3	<--------	
Signal earth	AB	102	7	N/A	
Screen earth	AA	101	1	N/A	
Group B: Control Signals					
Request to send	CA	105	4	-------->	
Clear to send	CB	106	5	<--------	
Data set ready	CC	107	6	<--------	
Data terminal ready	CD	108	20	-------->	
Ringing indicator	CE	125	22	<--------	
Carrier detect	CF	109	8	<--------	
Group C: Timing Signals					
Transmitter signal element timing	DA	114	15	-------->	
Transmitter signal element timing	DA	113	16	<--------	
Receiver signal element timing	DA	115	17	<--------	

The carrier detect (CF) indicates to the DTE that a carrier signal is being received by the DCE. The timing signals are used when synchronous communication is carried out. Either the timing signal is generated by a local device (usually a modem) and transmitted down the line to the far end (DA or DB), or the signal arrives from the distant transmitter (DD). In all cases the timing signals are square waves with equal ON and OFF periods and with the transition taking place at the centre of each data bit being transmitted.

The way in which these basic line signals are used in the transmission of a message is seen in an actual example. Consider the procedure required for the asynchronous transmission of a set of data from a DTE to another DTE using a pair of modems and the PSTN. We will use half-duplex operation to illustrate this, i.e. transmissions in one direction only, although in practice the procedure is likely to be simplified by full duplex working, i.e. transmission in both directions, which enables the transmit and receive functions to take place simultaneously. One simplification obtained is that some of the control signals (e.g. request to send and clear to send) can then remain in the ON state as long as the modems are switched on.

Transmission commences with manual or automatic dialling of the called telephone number. The dialled digits are carried over the PSTN and, in addition to the expected result of bell ringing and handset lifting, will turn ON (i.e. produce a logical 1) the ring indicator (CE) and request to send (CA) lines. (It is assumed that the data terminal (CD) line will already be ON.) The CA signal does two things: it transmits a carrier tone over the PSTN and, after a delay, turns ON the clear to send (CB) line. The carrier tone initiates the carrier detect (CF) line at the calling end. It also turns ON the CD and CC (data set ready) lines. In the meantime, after the delay, the transmitting data (BA) line is activated and a short 'request to send' message is passed over the PSTN to inform the calling end that all is in readiness for data transmission. The CA and CB lines at the called end are turned OFF, and with them the carrier tone from the called end. This sequence of events is shown in Fig. 4.8.

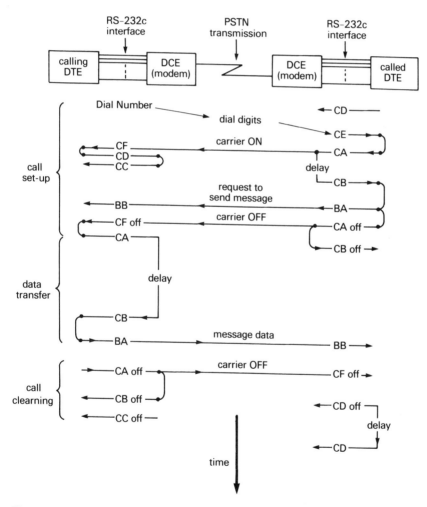

Fig. 4.8 RS–232C operation.

After a short delay at the calling end, prompted by the CA signal, the CB and BA lines are turned ON and the message data transmitted over the PSTN. This message data, of course, become the received data (BB) at the called end. At the cessation of the message data the CA line and the CB line at the calling end are turned OFF and the carrier tone across the PSTN is removed. This is noted at the called end by the CF logic, and CD is first switched OFF and, after a short delay, turned ON again in readiness for the next communication. In the meantime the CB and CC lines at the calling end are turned OFF or **cleared**, ready for the next communication request. The modems at each end are now set in the passive or waiting state ready for the next call.

4.5.2 The null modem

The example given above is a fairly complex one involving modems and the PSTN. Many applications of the RS-232c interface are simpler and may involve, for example, connecting a serial printer or other non-intelligent device to a computer. The distance between the devices may be so small that DCE controlling equipment will be unnecessary. In this case we need to arrange things so that two DTEs can signal directly to each other. Of course one of them can be rewired to 'look like' a DCE, but a more general solution is to insert a simple interconnection box referred to as a **null modem** between them (Fig. 4.9). Reasons for the particular cross-connections used will be apparent from our earlier discussion.

4.5.3 Synchronous transmission

When synchronous transmission is required, the clock signal is passed from the modem or other DCE or the DTE invoking the transmit signal element timing lines (DA *to* the DCE and DB *from* the DCE) depending on which device controls the timing.

Timing information is provided at all times when the carrier detect (CF) is ON. The timing information is received on a separate receiver timing line (DD).

4.5.4 X21 interface

We have seen how the RS-232c interface is applied to a telephone transmission system using a modem. The procedure becomes clumsy when applied to a purely digital or a **packet-switching** system. Packet switching is a method of transmitting a digital message as a series of short numbered frames, called **packets** across a communications system. Why this is done and the advantages thereby obtained are described in the next chapter. For the

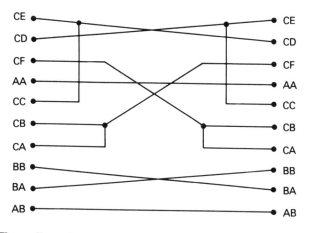

Fig. 4.9   The null modem.

present we will consider the DTE to be a device capable of accepting and transmitting data in this way, rather than as a continuous stream of bits. Because of the nature of the transmitted data, different protocol arrangements are required. One of these is the X21 interface, which forms the physical level interface for the X25 packet-switching system used by British Telecom and other PTTs (legal authorities for communication services in a given country). Its designated function is to provide a full duplex bit serial synchronous transmission path between the DTE and the DCE supplied by the local PTT for an all-digital PSTN or private network. A second standard interface protocol, X21bis, has also been defined for an analogue circuit-switching network and is, in fact, a subset of RS-232c/V24 but having fewer facilities.

X21 protocol needs fewer facilities than with RS-232c since it is only concerned with set-up and clearing operations associated with each call. The control of data transfer is the responsibility of other and higher protocols carried out in local computing equipment, as we shall see later. These limited number of circuit definitions for X21 are shown in Table 4.2.

A typical data interchange involving X21 and the higher layers accessed through the DTE is illustrated in Fig. 4.10. The DTE initiating the call first sets its control (C) level ON and at the same time its transmit (T) level to logical 0 (the 'waiting' state is logical 1). The DCE responds by transmitting a SYN character (a particular series of bits required by the higher protocol layers managing the data interchange) on the receive (R) line and follows this with a series of '+' characters. On receiving these the DTE itself transmits a data packet consisting of a SYN synchronizing character followed by the network address of the called DTE and a parity bit. Note that unlike the RS-232c information transfer, the X21 data exchange involves more than the transfer of a stream of digital bits between the two devices. Apart from the C

Table 4.2

X21 interface signal

	Function	Direction	
		DTE	DCE
Group A: Data signals			
Signal earth (G)		N/A	
DTE common return (Ga)		------->	
DCE common return (Gb)		<-------	
Transmit (T)	to convey both user data and control depending on the state of C & I	------->	
Receive (R)	as T in other direction	<-------	
Group B: Control signals			
Control (C)	provides control information to DCE	------->	
Indicator (I)	provides indicators to DTE	<-------	
Group C: Timing signals			
Signal element timing (S)	for bit timing	<-------	
Byte timing (B)	for byte timing	<-------	

and I lines the other transmission lines convey frames of data or control information, including synchronizing and parity information, which are acted upon by the higher protocol levels (such as HDLC or X25) with which X21 acts as the lowest level of communication.

The called DTE indicates its acceptance by turning its control (C) circuit ON and the DCE passes set-up information to the DTE. A 'ready for data' signal ensures that both ends of the link have their indication (I) circuits set to ON. Full duplex transfer of information can then take place across the link under the control of higher protocol layers in the DTEs. When the data transfer is complete the control (C) line from the called DTE is turned OFF (although the other end could also initiate this clear phase). A 'clear signal' request signal is passed across the link and turns the indicator (I) and control (C) lines OFF and both ends revert the transmit line (T) level to logical 1, resuming the waiting state at both ends for further transmissions.

4.6 SUMMARY Chapter 4 Efficient data transmission involves sending the information as a sequence of subsets of data, known as frames, across a transmission link. This technique enables data synchronization at both the channel and bit level to be

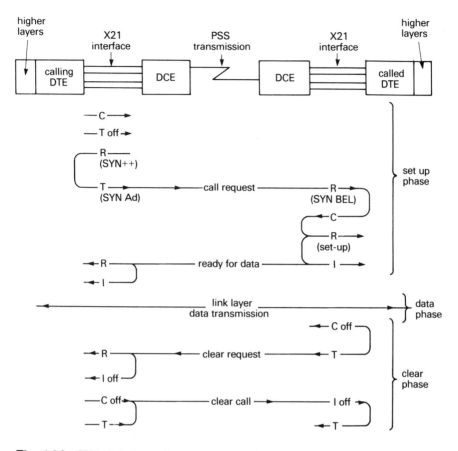

Fig. 4.10   X21 data interchange.

achieved and also provides a suitable structure for the application of error checking procedures. Analogue data for transmission are first converted into digital form using a process of sampling and quantization into discrete levels. Each level can then be described by a unique digital sequence. The process is known as pulse code modulation. Encoding the digital signal for transmission employs a number of alternative encoding schemes, known as source codes, each of which meets certain transmission requirements. Actual transmission of this encoded information requires the establishment of given interface conditions between equipment at either end of the link to control the transmission procedures across the link. These fundamental processes for simple data transmission across a link require considerable cooperative action between the two ends of the link. This is arranged by applying a set of agreed procedures, known as the physical level protocol, which are carried out through an interface connection. This connection can be complex, and to achieve the cooperation required between both ends has been standardized into a number of forms, each suitable for a particular range of transmitted information and equipment.

4.7 PROBLEMS Chapter 4

P4.1 A message 1101011 is to be sent with additional error correcting digits to enable cyclic redundancy coding (CRC) to be employed. The divisor word selected for transmission is 11001.

a Calculate the bit structure for the actual frame transmitted with the error control bits added.

b Prove that no error results if this frame is decoded at the receiver using a knowledge of the divisor word.

c Assume an error burst of three digital 1s is added to the transmitted frame at the least significant end. Show how the error may be detected using CRC.

P4.2 a Discuss the advantages of non-uniform quantization in a PCM system. What is the 'companding advantage'?

b Plot a continuous A-Law encoding scheme in the positive quadrant only for $A = 50$ and show how this can be approximated by seven successive straight-line segments where the slope changes by a factor of 2 at the junction between each pair of segments.

c What is the companding advantage, given that each segment covers a range of 16 non-linear coding decision values?

P4.3 Compare bit-oriented and character-oriented data link protocols in terms of advantages and disadvantages.

P4.4 It is not possible to sample a continuous (analogue) signal to derive an infinitely narrow slice of the signal. The sampling mechanism is such that the sampling pulse always has a finite width. Discuss the difficulties this presents to the digitizing process and suggest two ways of minimizing this effect.

P.4.5 Consider a signal which is comprised of three sinusoidal terms:

$$x(t) = 3 \cos 500t + 10 \cos 700t + 14 \cos 1000t$$

This is sampled in accordance with the sampling theorem and then quantized to 10 binary bits to form a PCM signal for transmission.

Find:
a The sampling rate required.
b The value of the quantizing interval level.
c The signal-to-quantisation noise ratio.

P4.6 If a binary signal is sent over a 3 kHz channel whose signal-to-noise ratio is 25 dB what is the maximum data rate possible?

P4.7 What is the function of a null modem? Give a diagram showing the interconnections used in a null modem and explain their use.

5

DATA SWITCHING

Transmission of data in any kind of network means essentially transferring or **switching** the data from one node in the network to the next in its passage from the originating station, attached to a node, to the intended receiving station associated with a different node. The process is called **data switching** and in this chapter we consider various ways of achieving this and look at the kind of problems that it generates.

So far, only the encoding and transmission of data between two devices over a single communication link has been discussed. In a switched network situation, such as the PSTN or a wide area network, the two devices may be separated by very many intermediate nodes. In general terms, two ways of doing this are possible. The intermediate nodes can be linked together to provide one particular route through the network out of a number of alternative possible routes existing between the two devices. Alternatively the data can be switched from node to node across the network with each internode link existing for a period long enough to transmit the message between the pair of linked nodes. In either case a choice, or series of consecutive choices, has to be made which we refer to as **routing**. The situation is simpler with a local area network which generally consists of a series of linked **nodes** having a fixed spatial relationship between them, and no alternative paths through the network (except that of direction) exist.

The nodes themselves are not concerned with the content of data being conveyed. Their purpose is to provide a **switching facility** that will transfer the data from node to node until they reach their destination. Each **end device** (e.g. computer, printer, terminal etc.) acting as the transmitter or receiver of data, will be attached to a given node. Sometimes the device attached nodes will be located at each end of a network with intermediate nodes free of the task of communicating directly with end devices. More usually, many nodes on the network will have attached devices so that the capability of transmission from or to a multiplicity of devices will exist. Further, certain nodes may be connected to nodes forming part of another, quite separate, network. In this case these nodes are called by a special name of **gateways**. They will need to provide a number of additional features since the method of operation of one network may differ considerably from the other network to which the gateway is connected.

Whatever the function of the nodes comprising the network we need to distinguish the way in which data are switched through them since the method of switching must remain constant throughout the entire network if the nodes are to function in complete cooperation, one with another.

Three switching methods are in common use:

a circuit switching,

b message switching, and

c packet switching.

5.1 CIRCUIT SWITCHING

This is a method we are familiar with in the operation of the public telephone service. In such a service three distinct phases of operation can be distinguished, namely: call initiation by lifting the receiver and dialling, conversation exchange and termination of the call by replacing the receiver.

In circuit switching these processes are formalized as:

a Circuit establishment. The setting up of an end-to-end circuit through an appropriate set of nodes located between the two end devices on the network, i.e. establishing a dedicated route.

b Data transfer. Transmission of data across the network following the dedicated route.

c Circuit disconnect. Termination of the connection after data transfer is completed and the separate node paths made available for other data signals.

The essential feature of circuit switching lies in the existence of a **dedicated communications path** between the two end devices. Circuit establishment demands a certain amount of elapsed time (**set-up time**), before a message can be sent. To do this a call request signal needs to be sent and this requires an acknowledgement. Further, each node needs time to set up the route of its onward connections through the network. As an example, the set-up time for the PSTN is relatively long (tens of seconds) but with a computer-controlled exchange this would be reduced to a few tens of milliseconds.

These three processes of circuit establishment, data transfer and circuit disconnect are illustrated by reference to the mesh network shown in Fig. 5.1. End device A requires connection to end device B. (A might be, for example, a microprocessor and B a printer.) First it is necessary to **establish** the complete circuit between them before data can flow. To do this, A sends a request to its connecting node 1 for onward connection to B. Stored in the node software will be details of routes available and their order of preference based on economy of usage and dynamic considerations such as channel availability. We will assume that node 3 forms the next part of the route (although it would be equally valid to choose node 2 or a route via 4, 6 or 4, 6, 3). A dedicated channel is allocated between nodes 1 and 3 and a similar request procedure is enacted at node 3 in which all onward routes are considered *except* that leading back to node 1. Assuming node 5 is selected, it remains for this to request access to its connected device B. If this is ready to accept data then an acknowledgement is passed through the dedicated route to device A and the process of **data transfer** commences. The allocated connection between A and B will almost certainly be duplex (i.e. signals/data can flow in either direction) so that after data transfer is completed either end can terminate the process. A reverse operation of signalling to each of the nodes in the connection path takes place in which their dedicated channel connections are **de-allocated**, thus making them available for other network messages. Note that once the completed circuit connection is set up, the data constituting the message can be sent as a single block and the only delay experienced will be **propagation delay** through the network. Since the intervening nodes are unlikely to have much (if any) data storage capacity, data flow must take place at a **constant rate** and this must be the same for devices at each end of the connection.

Circuit switching can be inefficient. For very short interactions it carries a

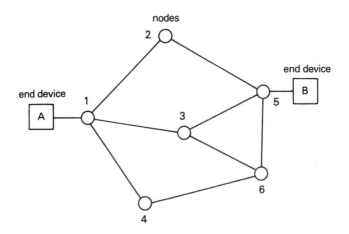

Fig. 5.1 A mesh network.

significant overhead in setting up the connection which could not be tolerated, for example, in bank cash transactions or airline booking information. In addition, channel capacity is allocated for the duration of the requested connection, even if little or no data are being transferred. It is, however, fairly well suited to speech transmission and finds its major use in the PSTN and in digital PABX (private exchanges) which may carry speech, data, facsimile and other services in a digitally encoded form.

Some improvement in efficiency is possible with private connection or **leased line** since the route can be a **dedicated** one and not subject to the setting up and disconnection process which occurs frequently in a public network.

5.2   MESSAGE SWITCHING

Message switching is rather more efficient. It is not necessary to establish in advance a dedicated path between two end devices. Instead, if a device wishes to send a message, it appends a **destination address** to the message data, indicating the end device (or devices) to which the message is to be delivered.

This message is then passed through the network from node to node over a convenient route existing at the time. At each node the entire message is stored briefly and transmitted to the next node, where the process repeats until the message has reached its destination. This is known as a **store and forward** technique[1]. Unlike circuit switching, it does not require a dedicated channel or elaborate set-up procedure but the entire message must be received at each node before it can be retransmitted. Thus, the total transmission delay is variable and generally longer than with circuit switching. This makes it particularly unsuitable for interactive traffic. It is an exceedingly reliable system, however, since the message is stored and checked at each intermediate switching node in its passage from transmitting to receiving end devices.

Referring again to Fig. 5.1, under message-switching conditions a destination address for end device B is appended to the message before it leaves end device A. This is noted by node 1 where the message and address are stored temporarily. Again, node 1 will consult its internal stored routing information and when a suitable route to the next node is available the data are transferred to it without waiting for any acknowledgement and without permanently allocating a channel to the selected node. It can do this since each node can choose its own time when to make the transfer of its stored data. Once transferred and the data received has been verified, the temporary memory locations are cleared and immediately made available for other incoming data. A similar process takes place at other nodes on the route from A to B until all the data have been delivered. No final disconnection of nodes in the network path is necessary. The technique is valuable when used for the transmission of a series of short messages which do not require acknowledgement. The major disadvantage is the variable delay that can occur. Note that the data need not be transferred at a constant speed throughout the network since the entire message is available for transmission at the rate demanded by the receiving node. Advantage is seldom taken of this for the very long messages involved. It is, however, useful for specialized 'mini-message' data which operate under packet-switching conditions considered next.

5.3 PACKET SWITCHING

In packet switching the nodes handle much smaller data lengths than are found in message systems. The message is divided before transmission into a series of sections of data called **data packets** having a maximum length of only a few thousand bits[2]. This technique has a number of advantages. First, the short packets experience minimum delay in progress through the network. The method used is still a store-and-forward process, but since the packets are small they are quickly copied by each node and require little memory space. Second, by appending a **sequence number** to each packet as well as its destination address, the nodes are able to interleave packets from several different sources and this leads to more efficient use of the transmission media. Fig. 5.2 illustrates how this interleaving can function.

Packet switching was first applied to wide area networks such as Arpanet and Tymnet[3] and these systems have set a pattern for later development. Two approaches are applied to the way in which this stream of mixed packets is handled by the nodes. These concepts are known as the **datagram** and the **virtual circuit**.

5.3.1 The datagram

In the **datagram** approach, each packet is treated independently and its route across a series of nodes to its destination can vary, dependent on the network traffic. Datagrams are single-packet messages, each of which contains the **full address** of the destination on its header and a **generator sequence number**. In many ways we can regard a datagram as a message system with the message reduced to a single packet.

It is possible that the packets may arrive in a different order to that generated due to a more favourable path being available for some packets at the time they are due to be transmitted. This was shown in Fig. 5.2. Since each packet carries the destination address and its generation sequence number the packets can be reshuffled into the correct order without too much difficulty, providing adequate storage capacity exists at the receiver end.

An example of a datagram network is the Canadian DATAPAC system[7]. DATAPAC was an early system installed in the mid-1970s. The network offers a call-based service with two types of call: permanent and switched.

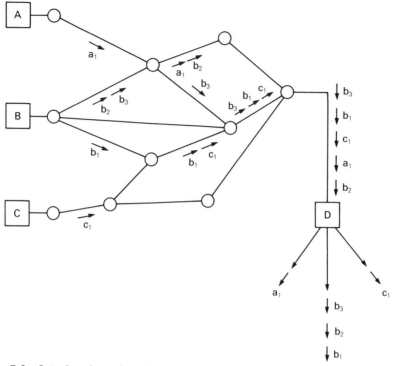

Fig. 5.2 Interleaving of packets in a network.

Permanent calls are analogous to a private telephone line and require no call set up or call clearing. Switched calls are analogous to public calls and do require prior set up and subsequent switching. It is also possible to designate packets as 'normal' and 'priority' with the latter given preference for onward transmission. The size of a packet is 128 bytes for priority service and 256 bytes for normal service.

Whilst a dedicated datagram service was valuable in the early days of packet switching, it was subsequently found possible to provide a datagram service as a subset of a virtual circuit service just as economically. Consequently, few networks are now based solely on the datagram concept. One of the exceptions is where the business of the network is entirely concerned with short messages of one or only a few packets — a good example is electronic funds transfer transactions between a supermarket and a bank, where the overhead in setting up a virtual call is generally not worth while.

5.3.2 The virtual circuit

The **virtual circuit** approach is quite different. Here a **logical connection** across the network is established before the packet is sent, i.e. a specific ordered series of nodes is determined through which the packet may be transmitted to reach the end device. This permits the packets to be sent as a complete stream through a defined path and each node does not need to make a routing decision for each packet received. Also, it becomes possible for the packets to be delivered in their order of generation if this is desired. Note that this does *not* mean that there is a dedicated path from source to destination as in circuit switching. A packet is still stored temporarily at each node and queued for onward transmission together with other packets in the holding queue. Only the route is dedicated to the message and not the lines themselves. Access to the data is obtained using a set of **numbered logical channels**. Initially, when a virtual circuit is being set up, a logical channel not in current use is assigned at each end of the transmission link. Each subsequent packet routed through that virtual circuit has the logical channel number of the sending and receiving ends included in the information at the beginning of the message — the **header**. This enables a long message to be broken down into a large number of packets before transmission and correctly reassembled at its destination.

As in circuit switching there are the two possibilities for the virtual circuit so formed. Either the virtual connection can be **permanent** and thus equivalent to a leased line, or the virtual connection can be defined only for the duration of the message — the equivalent of a dialled call. This latter is known as a **virtual call** and is initiated by means of a **call request** packet. This call request packet must contain both the called end device number (or address) and the number of a free local logical channel. If the call request is valid, the network routes it to the node connected to the end device called. If this device is free to receive the call, the node inserts a number of a free local logical channel which can carry the incoming data and forwards the call request as an **incoming call** packet. The end device then returns a **call accepted** packet, which is routed back through the network to the node connected to the sending device. This node then sends a **call connected** packet to the called device node and the establishment of the virtual call is completed. The transmission of these various controlling packets is very rapid since the control packets themselves are quite small and, by carrying out a rigorous 'hand-shaking' protocol of this kind, the routed call is able to be set up unambiguously between the two devices. This packet equivalent to the switched circuit is, like the switched circuit, more efficient where two end devices wish to exchange information over an extended period of time.

5.4 EVENT TIMING AND SPEED CHANGING

The comparative behaviour of these different forms of switching techniques in terms of the delays experienced during set up and transmission is shown in Fig. 5.3.

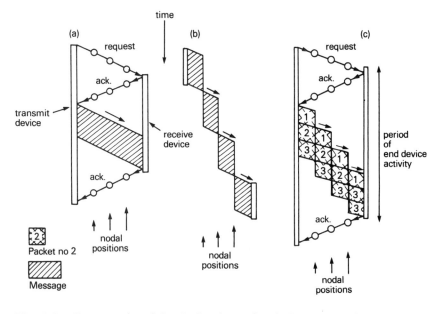

Fig. 5.3 Comparative delay behaviour of switched networks.
(a) Circuit switching; (b) message switching; (c) packet switching.

The delays shown are: **propagation delay**, which is the time taken to transmit the data to the next node, **node delay**, i.e. the time taken by the node to carry out the switching process, and finally, **transmission delay**, which is the time required for the sending end device to transmit the data across the network. This latter will depend, of course, on the length of the message and speed of transmission through the network.

With circuit switching (Fig. 5.3a), the set-up procedure involves all three delays for the request and acknowledgement information that completes a round trip through the network back to the sender. This is the **set-up delay**, a principal component of which is the node delay that occurs at each node as it establishes a route for the connection to the next node. Following the connection, the message is sent as a single block incurring a transmission delay for the message. A further delay is experienced when the message is acknowledged. Message switching (Fig. 5.3b), does not include a call set-up time. Node delay, is however, involved since each node must receive all the message before it can begin to transmit this to the next node. Packet switching (Fig. 5.3c), will experience rather longer delays than circuit switching which operates in a similar manner. An increased value of node delay occurs, since each packet is stored to await onward transmission through the network.

We mentioned earlier that one of the purposes of packet switching is to improve the utilization of available transmission paths. Another important advantage is the facility for **speed changing** inherent in the store-and-forward transmission technique. In a digital communication path, the data rate of the line is strictly irrelevant to the information it carries. With packet switching using store-and-forward techniques, any terminal or node is able to choose the data rate appropriate to its traffic and its method of operation. In a circuit-switched system this is not possible.

5.5. NETWORK CONTROL

Control for data flowing through a switched network is vested in the nodes, which are generally intelligent microprocessors, i.e. devices with memory and stored program/data. Three particular areas of control operation are carried out at node level. These are:

a routing,

b flow control, and

c error control.

5.6 ROUTING

The difficulties and realizable efficiency in routing data through a network depends very much on the layout of interconnections — the **topology** of the network. For example, if this is arranged in the form of a star with connections made only to a control hub or exchange then only this control node need be in possession of information defining the network topology and a routing procedure or **algorithm** is simple to devise. As mentioned earlier, the problem also presents no difficulty with local area networks where the nodes are arranged in a line (bus) or a ring.

The real complication is present when we consider an **irregular mesh** network such as we find with the PSTN and in wide area networks. A method in wide use is to make decisions on the route for onward transmission based on a **routing table**. This is illustrated in Fig. 5.4. A packet enters the network with a destination addressed appended to it. Routing decisions are made individually at each node by reference to the table which indicates a preferred next destination based on the network topology and other considerations. The table also includes alternative routes which will be chosen in the event that the preferred route is congested or not available.

The diagram shows routing tables located at two nodes, A and B, each containing a single alternative route in case the first one cannot be used. Note that the routing tables take no account of the earlier history of the packet, i.e.

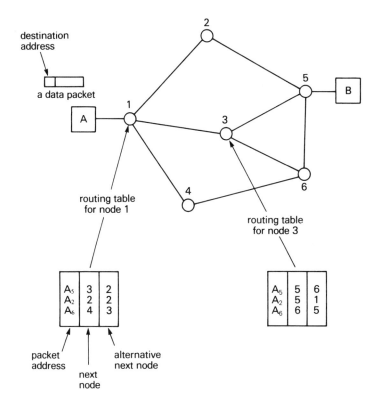

Fig. 5.4 Use of a routing table.

the number of nodes through which the packet has previously travelled. It is only concerned with forward information.

Routing design methods to achieve optimum results can be complex. Several mathematical models have been suggested which make use of **queueing theory**[2] and other analytical methods. Use has also been made of **simulation techniques** which can be validated on an actual network[3].

The method shown in Fig. 5.4. is known as **deterministic** or **fixed routing**. Here the message routes are generally unaffected by variation in network traffic. Since they are never changed or changed infrequently they can be allocated centrally. Advantage can be taken of this by assigning a route only on a user log-in basis. The route is assigned depending on the network traffic prevailing at the time and does not change for the duration of the users' logged-in period. This technique is applied in the Tymnet network[4].

An improved scheme is **adaptive routing** where use is made of dynamic information on the traffic flow which is applied to modify the fixed tables. The information is derived from the traffic density measured at different exit ports on neighbouring nodes. It is difficult to make use of forward information available at distant nodes in remote parts of the network because of the problem of transmitting the data over the network itself. As a consequence, adaptive routing is only partially effective. It is, however, a useful method and various forms of it have been applied. We look first at a method which makes use only of local information.

5.6.1 Isolated adaptive routing

Here routing decisions are taken only on the basis of information available locally in each node. This will consist of:

a   a pre-loaded routing table, as in fixed routing,

b   the current state of the on-going connection (i.e. free or busy), and

c   lengths of the packet queues awaiting the use of each node.

The routing algorithm is programmed to make a choice between alternative routes and expressed through a calculation based on queue lengths and knowledge of the network topology giving a **bias** towards the choice of better onward connections[5]. This information is applied to modify the initial routing table held within the node. An example is shown in Fig. 5.5.

A node is considered to have available a **primary** and a **secondary route** to

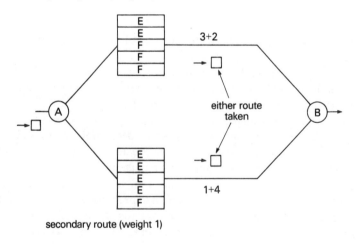

Fig. 5.5   Isolated adaptive routing.

the next node as indicated in the diagram. The number of **free spaces** currently available on the **ouput queue** for the primary route is shown here as 2 and for the secondary route 4. With the weighting previously worked out from the routing table, the routing process will choose between primary and secondary routes having a ratio of 3 + 2 and 1 + 4, i.e. 5 : 5 giving an **equal probability** of using either route. However, if, on the other hand, the number of empty queue slots is, respectively, 4 and 1 then the probability ratio will be 3 + 4 and 1 + 1 giving 7 : 2, so that the bias will be towards the **primary route**. If one of the output queues is full then the packet is sent to the other route irrespective of the number of empty slots indicated. If both output queues are full, then the packet is discarded to be transmitted later over the network. A suitable request for retransmission will then have to be made back over the network to the originating source.

5.6.2 Distributed adaptive routing

A different routing algorithm is implemented in Arpanet which is also adaptive to local conditions. This is known as **distributed adaptive routing**. The object is to find paths of least delay for the network traffic. Here the average delay is measured at each node for every outgoing link to its neighbouring nodes. This is carried out every 10 s and the information so obtained is propagated from the node to other nodes in the system. Each node can use these estimates to compute its fastest path to ongoing nodes and, in effect, provide a dynamical data base of 'shortest times' for its local section of the network topology. These data can then be used to obtain a decision on routing for all packets received, knowing the packet destination addresses. The delay incurred by the updating information packet circulated around the network can be considerable, although independent of the total size of the network. In an earlier version of Arpanet the update interval was only 0·67 s and circulating and updating the delay table accounted for nearly 50% of the bandwidth available for some lines. The interval between updating is now longer and in addition, delay information is only transmitted when there is a significant change detected in the information to be circulated from the last update table. In a practical operation this modified scheme does not add too much to the network overheads.

5.6.3 Flooding and random routing

It is possible to dispense with a formal systematic routing algorithm altogether, using the techniques of **flooding** or **random routing**. These methods are particularly useful in military communications where the configuration may not be known exactly and is subject to modification by drastic enemy action.

With **flooding**, multiple copies of the packet are forwarded to all onward nodes from the node concerned — except the one on which the packet was received. A previous calculation needs to be made of the **shortest number of nodes** to be traversed to reach the destination. This is known as the **hop count**. When this limit (or just beyond it) is reached, the radiating packet is destroyed so that only the packet reaching the destination by the shortest route is preserved.

With **random routing**, a random choice is made of an output path (other than the arrival path) along which to direct the packet. This happens at the next node and so on. The method has been likened to a **random walk** since the packets do not progress systematically towards their required destination. Again, the **hop count** is used as a measure to decide upon the elimination of a packet at a given node.

Neither of these routing methods requires information about the network topology and, in the case of flooding, the packet is guaranteed to reach the destination if this is at all realizable.

5.7 FLOW CONTROL

The second feature of packet transmission in a mesh network which requires consideration is that of **flow control**. Flow control means the methods used to keep network data moving. It is not enough to maintain a sufficiently high data flow within the network itself, since the network is not able to deliver packets faster than they are being received and this alone can cause a bottleneck. It *must* be possible for the network to **signal back** to the sender a requirement to suspend packet delivery for a period of time if this becomes necessary. In extreme cases, part of the network may become overfilled with packets so that it can become impossible for packets to move around the network at all. This is called **congestion** and has close analogies with a similar problem in an automobile traffic situation.

Flow control methods are generally based on the acknowledgement of packets transmitted by the source and received at their destination. Acknowledgement can be **positive** (ACK), indicating that the packet has been accepted correctly, or **negative** (NAK) indicating incorrect reception or the loss of a packet with consequent need for retransmission. The flow control protocol in use must arrange that the node connected to the source of data does not accept packets at a rate greater than the end device can accept them. One method is to continually monitor the length of the packet queue at each node and to forward this information to its immediate neighbour which modifies its routing table accordingly. It is also necessary to ensure that the total number of packets in transit does not exceed the total storage capacity of the network.

5.7.1 Stop-and-wait flow control

The simplest form of flow control is known as **stop-and-wait**. The receiving device agrees to accept data by responding to a request transmitted across the network. The sending device then transmits its data. After receiving these the receiver must again agree to accept a further section of data before this is transmitted. The method works well where the message is in the form of a long block of data. It meets with problems in the packet situation where the data are split up into short packets for transmission. The difficulties are connected with effective utilization of the transmission channel and the result is to extend the propagation delay and reduce the overall rate of transmission. We can understand this if we look at an example, illustrated in Fig. 5.6.

Packets are sent one at a time with a pause between them to permit the inclusion of an acknowledgement signal. Node A sends a packet n_o to node B which responds with a short ACK packet if the packet is received correctly. Packet n_1 is then transmitted (Fig. 5.6a). If the packet is imperfectly received then this is detected through an error detection mechanism contained within the node software and a NAK is returned to node A. Since this maintains a copy of the previous packet, pending the receipt of a successful acknowlegement, it is able to repeat the transmission and again wait for an acknowledgement (Fig. 5.6b). With this control scheme there is no need to distinguish between packets. The successful receipt of an ACK packet initiates the transmission of the next data packet and so on.

In a practical situation, the short acknowledgement packets may themselves be corrupted by noise, or fail to arrive so that the initiation of further data transfer is held up. A method of overcoming this situation is referred to as **time-out** and involves the retransmission of the last packet sent if an acknowledgement (ACK or NAK) is not received before the expiry of a predetermined time interval.

Other difficulties remain, however, such as a correctly received data packet which is not properly acknowledged. This will result in a further transmission of a second copy of the packet. The flow control protocol must therefore be able to distinguish between new data packets and repeat copies. A simple method is to provide an **alternate bit** in the data packet header which takes up alternate values of 0 or 1 with each successively generated packet. A node having received a data packet with a 0 in this position would expect the next packet in the sequence to contain a 1 and the next a 0 and so on[6]. It is usual to include the alternate bit in the ACK packet as well, to provide a second check.

In may situations, the data traffic will constitute an exchange, so that data packets and control packets will be travelling in both directions through the network. This gives an opportunity to add the ACK signal to the header of the next packet to travel in the appropriate direction and this is often done. In the event that the traffic becomes unbalanced for a period, in which there are no data packets travelling in the reverse direction, then a special short acknowledgement packet is sent.

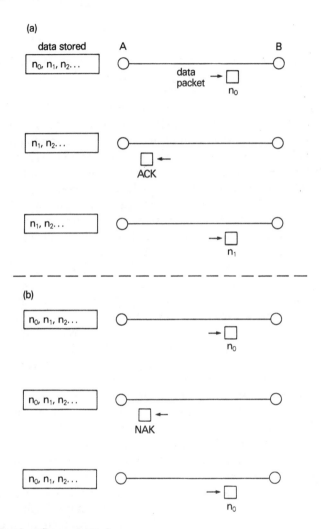

Fig. 5.6 Packet flow control.

5.7.2 The sliding window

The problem with stop-and-wait schemes is that only one packet or frame can be in transit at any one time. Where the number of bits that can be accommodated in a given link is greater than the number contained in a packet, then the method is inefficient. The solution is to allow multiple frames rather than a single frame to be in transit at a given time.

When we do this we need to make sure of their order of reception since some may be delayed (or even lost in transit). Hence each packet must carry a **sequence number** indicating its order of generation. It is then possible for the receiving node to indicate in its acknowledgement the sequence number of packets received. The number of packets that may be transmitted before an acknowledgement is received, is termed the **window width**. A multi-packet node transmission scheme employing sequence numbers and a varying window width can be a little complex, but the general operation of such schemes may be understood from the following example with reference to Fig. 5.7.

Initially, all sequence numbers are set at zero so that the first packet to travel from node A to node B will have sequence number s_0, the next expected number will be s_1 and so on. The window width, w, in this example, is set at 5 so that packets up to s_4 may be sent without acknowledgement. At the

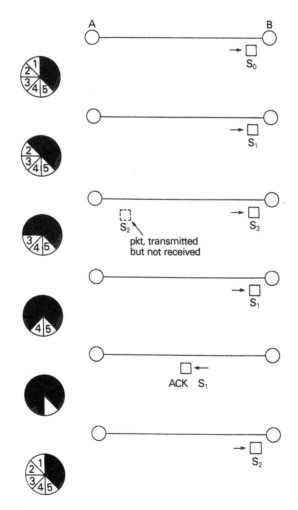

Fig. 5.7 Use of a sliding window.

receiving node the packets are checked for accuracy and the sequence numbers examined. They are expected to be received in correct order and if so, an acknowledgement to packet s_4 indicates that packets with sequence numbers 0 – 4 have *all* been successfully received. A further batch of packets can then be sent. The number of packets that can be sent from node A to node B depends on the size of the window and with each packet transmitted the size of the window is reduced by 1 until the acknowledgement of a multiple packet enables it to expand to its limit value of 5 again. This precaution is necessary to control the destination of packets that arrive out of order. Suppose after transmitting the first two packets bearing sequence numbers s_0 and s_1 the next arrival is s_3, then at node B this will not be acknowledged or stored. When final group acknowledgement is made this will consist of sequence number s_1 (not s_4) indicating correct reception up to and including the second packet — all subsequent packets being ignored. The transmitting end will delete its copies of packets 1 and 2 and commence its next batch transmission at packet 3 carrying sequence number s_2. If a packet or a group is repeated then this too will be ignored since it will not be carrying the correct sequence number. The next packet is likely to be in the expected sequence and will take the place of the repeat packet in the reception register of node B.

Of course, the acknowledgement packets themselves may be lost or delayed in transit. Accompanying each packet stored at a node concerned will be a time (the time-out) within which the packet must be acknowledged. Should this time elapse before the packet is acknowledged, then the node will retransmit the packet with its original sequence number. However, since packets can only be accepted in the order of transmission, all the later an-acknowledged packets must be transmitted as well.

This is, in fact, one of the difficulties of this particular 'sliding window' protocol and alternative, rather more complex methods, but still based on the basic window method, have been described in refs [7,8].

5.7.3 Isarithmic flow control

A flow control specifically designed to avoid congestion in a network is known as **isarithmic flow control**. As proposed by Davies *et al.*[7], the total number of packets in the network is maintained constant. The packets are of two kinds: data packets and empty packets. An empty packet is created as soon as a data packet leaves the network. A data packet can be accepted by the network following the location and erasure of an empty packet. The method is similar to the operation of a token bus or ring network to be considered in Chapter 7. Where no data packet is available for entry, then an empty packet is routed to a random destination node.

The performance of an isarithmic network has been determined through a series of simulation studies. It has been found that this is critically dependent on the total number of messages and nodes in the network. There is, in fact, an optimum number of packets per network node for minimum congestion. This is low and lies between $2N$ and $4N$ for a maximum node output queue size of 3 – 8 respectively, where N is the number of nodes in the network. As with stop-and-wait flow control, the overheads for sending empty packets can be minimized by appending these to data packets flowing in the opposite direction.

5.8 OPTIMIZATION

The calculation of optimum requirements for a packet-switched network is a complex process. It concerns choice of transmission method, topology, line capacity and allocation of data flow. Very often the optimization of a network commences after the method and topology have been chosen, since there may well be other requirements determining these choices which have more to do with installation economies and practice. Commencing with a consideration of line capacity and data flow optimization, there are some well-known guidelines available which can help in the solution of the optimization problem.

Line capacity can always be increased to reduce the average network delay which is often the single factor dominating the performance of a network. The limit to which this can be increased is determined by cost, so it is useful to know the relationship between the line capacity and performance in terms of delay so that the minimum capacity may be chosen.

If we consider a given channel, n, and a data flow of X packets per second, we can make the assumption that the network delay depends only on the channel flow. The calculation of an average packet delay, T, can then be made without resort to complex queueing calculations but at the cost of some simplifying assumptions. First, the assumption is made that with each line there will be an average queueing delay denoted as T_i. To apply this we use **Little's result** an equation first proved by D.C. Little in 1961, which, when applied to packet switching, states that the mean number of packets, N, in the network is the product of their mean arrival time, λ, in packets per second and the average packet delay time T, i.e.

$$N = \lambda T \qquad (5.1)$$

The number of packets contained in each channel queue is thus $\lambda_i T_i$ and summing over all the channels of the network gives:

$$\lambda T = \sum_{i=1}^{m} X_i T_i, \qquad (5.2)$$

where m is the number of lines (channels) in use.

Two further factors need to be introduced into this equation. These are the times required to service packets in a given queue, which is proportional to packet length, L, and some way of taking into account that different queues will have a different distribution of arrival times for the packet streams. A relatively simple solution has been suggested by Kleinrock [9] which, although not rigorous, appears to provide useful working results in most cases. He assumed that the packets arrived at a rate of X_i packets per second, in accordance with a Poisson distribution probability[9] and that the packet service times are exponential with a mean of a^{-1}s (i.e. the average packet length in bits is given by $1/a$). The average delay for a simple queue may then be stated as:

$$T_i = 1/(aC_i - X_i) \tag{5.3}$$

where C_i is the capacity of channel i in bits per second.

Substituting in eqn 5.2 produces an overall average queueing delay which we need to minimize for optimum working, viz:

$$T = X_i \sum_{i=1}^{m} \frac{X_i}{\lambda} \left\{ \frac{1}{(a\,C_i - X_i)} \right\} \tag{5.4}$$

This solution ignores the processing time for the packet and the propagation delay, as well as the time required to send acknowledgement and other non-productive packets. A more accurate formulation is give in ref [10].

5.9 LOGICAL, VIRTUAL AND TRANSPARENT FACILITIES

Three concepts which are in wide use when describing communication networks and their protocols are the following:

a virtual,

b transparent, and

c logical.

They have rather special and precise meanings in this connection and will be clarified in this section. The word **virtual** referring to computer facilities or to data indicates that the item in question appears to exist to the programmer but in reality does not exist in that form. An example would be **virtual memory** in which a programmer appears to have unlimited memory available to store his program, whereas in practice the memory is finite but has automatic and rapid access to a much larger secondary memory or backing store. Before the limit of the finite internal memory is reached, the computer arranges for data to be automatically interchanged with a section of the backing store, usually a fixed amount of so many kilobytes. The internal memory is now free to accept more data which may, in turn, be transferred automatically to the backing store if the size of the program requires it. This process is sometimes called **paging**. Similarly, a **virtual circuit** is a communications circuit which exists in terms of effective connection but where there is no such physical circuit to precisely carry out this connection. In reality, the circuit will be shared by several transmissions, each appearing to have sole access to the circuit media.

Network **transparency** indicates something which appears *not* to exist whereas it does; i.e. the user thinks he is directly connected to some service or end point whereas in reality his connection is via a complex set of software/hardware connections.

Logical connections are, in effect, make-believe direct connections in which the connection is made eventually to the service requested but goes through a number of logical devices (concentrators, controllers, front-end processors etc.) which the user does not need to be aware of.

The words **virtual** and **logical** are often used interchangeably; e.g. 'virtual terminal' and 'logical terminal'.

We can have logical and physical messages. Thus the **logical record** can be the output of an applications program, whereas the **physical record** is what is actually written onto disc and can contain additional information or a number of such logical records. A good example of this is a network protocol framework structure in which the actual logical message is itself quite a small part of the frame, the rest of which consists of control and other information (we consider several examples in Chapter 8).

SUMMARY
Chapter 5

Data transmission through a network requires a message-passing technique to be set up which is capable of operating throughout the network. The actual process of passing the message from node to node across the network is called **switching**. Three techniques that have evolved to achieve this are circuit switching, message switching and packet switching. These each have certain characteristics making them suitable for handling different kinds of messages and at different levels of efficiency. An important distinction is the length of the data section that may be transmitted in a single operation from node to node. Certain advantages in control and efficiency are apparent if the message is split into a series of smaller messages, known as packets of information. This method also is particularly suitable for digital transmission.

Whatever the method chosen, it becomes necessary to define an internodal route through a given network which takes into account the passage of other messages over the same network. Various routing methods are devised for the different switching schemes to avoid **congestion** of data traffic and to make best use of the network design or **topology**.

Control of traffic in the network also requires that the data be transferred completely without any missing sections and in the order of the original transmission. Methods of control to achieve this make use of acknowledgement signals which are sent from the receiving to the sending device/node. Since the transmission of these signals will themselves lengthen the time of delivery for a message across the network, a number of carefully worked out schemes have been devised to minimize the number of separate acknowledgements that are necessary without affecting the accuracy of transmission.

PROBLEMS
Chapter 5

P5.1 A circuit switched network has an average length of 500 km between a given pair of end stations and passes through 50 nodes across the network. Each node requires a set-up time of 20 ms. Assuming a data rate of 10^4 bps and a propagation velocity of 2×10^8 m/s, determine the total time required to transmit a message of length **a** 10^6 bits and **b** 10^3 bits. Comment on the efficiency of the process for the two cases. (Request and acknowledgement signals are assumed to be of negligible length.)

P5.2 Given that the efficiency of a packet-switched system is set by the ratio of the propagation time across the link divided by the message transmission time, obtain the minimum length of message (frame size) possible to reach an efficiency of 50% with a data rate of 4k bps and a propagation delay of 20 ms.

P5.3 A data packet transmission network has the following characteristics:

Request packet length = 10 bits
Acknowledgement packet length = 10 bits
Data packet length = 500 bits
Number of packets transmitted = 1000
Propagation delay = 100 ms
Node delay = 1 s
Transmission bit rate = 9.6 kbps

There are 50 nodes in the network which has a total length of 5 km.

Find the effective overall data rate for the network. (Assume request and acknowledgement packets take a negligible time to transmit.)

P5.4. **a** A three-link linear packet-switching network is shown in Fig. 5.8. Find the average delay over the network which has the following parameters:

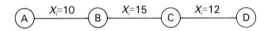

Fig. 5.8 Diagram for problem P5.4.

packets arrive at the network at an average rate of 50 per second;

each packet has an average length of 500 bits;

the capacity of the network is 9600 bps for all links.

b Deduce the minimum network capacity. Is this adequate?

P.5.5 Define the following terms:
a time-out;

b isarithmic flow control; and

c stop-and-wait flow control.

P5.6 A sending device transmits digital information at

a rate of 9.6 kbps. Information is transmitted in packets of length 960 bits with intervals of 500 ms between them. Assuming that a store-and-forward transmission technique is used, what is the minimum continuous transmission rate that can be used? What is the minimum storage required?

6

COMMUNICATION NETWORKS

We can distinguish two basic ways of conveying digital data across a communications network. The first is to **switch** the data from node to node across the network following a route selected out of a multiplicity of possible routes and with the selection either determined in advance or decided dynamically at each node receiving the data. This method is appropriate to a **wide area network** (WAN), i.e. a network designed to communicate information over large distances through very many intermediate nodes and generally operating in the form of a mesh network. Several techniques for data switching were discussed in the previous chapter. Note that not all the nodes in a network will be concerned in transmitting a given message — only those forming part of the transmission route for that particular message.

The situation is quite different in the second method, in which the data are **broadcast** through a network and where the choice of transmission route is limited to direction along a single route linking all the nodes in the network. Here *all* the nodes receive *all* the messages carried by the network. This method is appropriate to a **local area network** (LAN). This network may be defined as one serving a limited geographical area to provide interconnection between a variety of data communication devices.

Whereas a WAN is designed to carry data at a range of speeds between a few hundred bits per second to several thousand, the LAN is usually capable of transmitting data at several million bits per second. In this chapter we will be considering long-haul and other networks of the WAN type, leaving until the next chapter a consideration of local area networks.

6.1 NETWORK TOPOLOGY

The **topology** of a network refers to the way in which the nodes of a network are interconnected. Whilst the two categories of switched and broadcast networks serve to differentiate their primary purpose, there are a number of different forms of interconnection that may be used.

The most common topologies are:

a mesh,

b star,

c bus,

d tree, and

e ring,

with their interconnection arrangements shown in Fig. 6.1.

6.1.1 Mesh

Networks describing a mesh topology may be **fully-connected networks**, as shown in Fig. 6.2 or **random networks** such as the PSTN and other networks

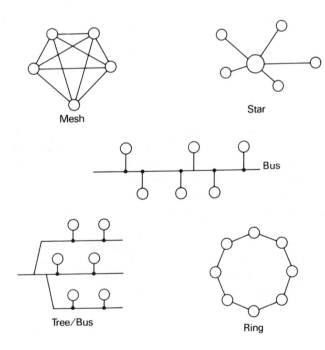

Fig. 6.1 Network topologies.

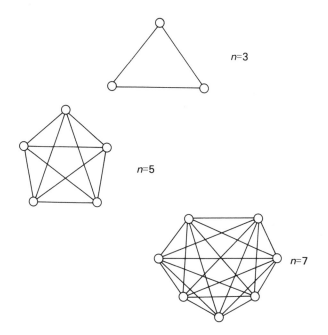

Fig. 6.2 Fully-connected networks.

illustrated previously in Chapter 5. In a fully-connected network, separate bi-directional links are established between each pair of nodes. No intermediate nodes are involved and each end device is linked directly via its communicating node to the node of every other end device. This carries with it a heavy penalty in terms of numbers of interconnections required. As we see from Fig. 6.2, a fully-connected network having n nodes will require $n(n-1)/2$ links. Thus the complexity of the network will increase proportional to the square of the number of interconnected nodes. Not only is this uneconomic in terms of line usage, since most of the links will be idle much of the time, but the addition of each new node to an existing network having n nodes will mean the installation of a further n lines so that even with a modest network, any expansion will prove an expensive operation.

The solution, as indicated in the previous chapter, is to introduce the task of line selection at node level by means of a switching technique. This means, of course, that since the lines are somehow shared between the available nodes then a direct connection between two given end devices is no longer possible and the message needs to be conveyed via a series of intermediate nodes. In the PSTN this function of node switching is carried out at the branch or trunk **exchanges**. In network terminology we refer to this complex mesh topology as connection through a series of interconnected **star networks**.

6.1.2 Star

In a **star topology** a special central node or **hub node** is added, to which all other nodes are connected. Any node can communicate with any other node via the central node or **switching centre**. Whilst this implies the installation of a complex central node, the communications processing burden on the other nodes is minimal and this makes the system economic for **terminal intensive** requirements.

It is currently a very important form of network for digital PABX or the **computerised branch exchange** (CBX). The modern CBX will use packet switching and the system may be arranged to carry several different services simultaneously, examples being speech, data, facsimile and slow-scan television. This integrated transmission of a number of different services over a common network will be considered in more detail in Chapter 9.

Since all the interconnection logic is incorporated in a central switching exchange, it becomes possible to carry out a number of other functions in the hub node. For example, the hub node can contain memory storage such that conversion of the data transmission speeds of the sender to that required by the receiver may be carried out. The sending and destination end devices may also operate using different communication protocols and character sets. Thus the hub node can act as a protocol converter or gateway. In fact, one of the most significant features of a star network is that much of the intelligence required to control the network can reside in one place — the hub. This is why the system is of particular value for simple terminal connection to devices which have little internal intelligence. The star network is, however, vulnerable to failures of the hub node and can be costly to install since the cabling alone can represent a considerable capital investment for the network.

6.1.3 Bus

In **bus topology** the communications network is simply the transmission medium itself. All the nodes are connected directly to a linear transmission medium, termed the **bus**. A transmission from any node propagates simultaneously in both directions through the medium, thus providing bi-directional transmission to all other nodes and their connected end devices.

As in mesh topology more than one end device may communicate to the network simultaneously. However, in the case of the bus network there are no controlling nodes capable of intercepting the several messages and selecting, storing and routing them correctly to their different destinations. The network *is* the medium and special arrangements have to be made at the network nodes, here called **transceivers**, connected to the end devices (or sometimes in the end device itself), to ensure that simultaneous transmissions do not cause disruption to other transmitted messages. This problem is called **contention** and is a major operating difficulty with all bus systems. It arises because of the distributed control of the network in which control rests entirely with the end nodes or devices. These need not only to prevent contention taking place, but also to determine which end device is permitted to transmit at any given time given the prevailing level of network traffic. Thus complex operating protocols, referred to as **medium access protocols**, are incorporated, which function in conjunction with other nodes on the network to permit access only at times when uninterrupted transmission is possible. This will be discussed further in Chapter 8.

A second practical problem is associated with the use of a common medium where the distance between pairs of communicating devices can vary from a short link to a neighbouring node to the full bus length. This has to do with ensuring that the signal strength at the receiving node lies within an acceptable range for the receiving device to function correctly. In a simple **point-to-point** link such as that found in mesh and star networks, the characteristics of the transmission link and the distance apart of pairs of neighbouring nodes is known and it is a simple matter to **balance the line**; i.e. ensure that the received signal strength is correct and the frequency/phase distortion occurring on that particular length of line is corrected. In a bus network it is necessary to carry out this signal balancing for all the permutations of inter-node line lengths found in a network. For an n-node system then $n(n-1)$ signal adjustments would need to be carried out simultaneously. This would be difficult to carry out in practice and a solution is to consider a large bus network to be divided into segments and to carry out balancing within each segment, using, if necessary, additional transceivers between consecutive segments.

6.1.4 Tree

A distinction needs to be made between the tree-connected network shown in Fig. 6.3a, in which a hierarchical set of nodes are connected together, and a bus/tree topology, shown in Fig. 6.3b which may be considered as an extension of the bus topology. The former is to be found in certain types of WAN and is used mainly for the simplicity in routing control possible with this form of network interconnection. Since only one network path exists between each node pair, then all the information necessary can be contained within the routing address. All that is needed is information on how the message data are to be moved up or down the network.

The tree topology illustrated in Fig. 6.3b finds its main use in the LAN. As shown in the diagram, the transmission medium is a branching bus having no closed loops. In order to simplify interconnection between branches, no more than two nodes are permitted, therefore, between those repeaters linking the branches of a network. One difficulty with this arrangement is that the signal will be propagated in the different branches at different speeds and be reflected at the end of a branch in different ways unless the lines are perfectly

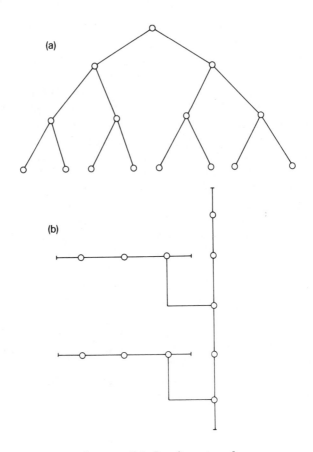

Fig. 6.3   (a) Tree topology.   (b) Bus/tree topology.

matched. For this reason, tree-connected bus networks are generally run at much lower speeds than those applied to a single bus network.

As with the bus system since all the nodes share a common transmission link, only one device can transmit at a time so that some form of **medium access control** is required to determine the order of transmission for the nodes. This generally takes the same form as for single bus networks.

6.1.5   Ring

The ring topology consists of a series of nodes connected to each other to form a closed loop. The nodes are arranged to accept data transmitted to it from an immediate neighbouring node to which it is connected and to forward the data bit by bit to its other connected node. It can only do this in one direction, however, so that the nodes, now termed **repeaters**, have a very simple task to perform. Contention problems such as those experienced in bus systems are absent, since messages are relayed sequentially through the ring. It is a fairly simple matter, therefore, to prevent a repeater from accepting data from its associated end device for transmission if the repeater is engaged in forwarding a message already present in the ring.

In a ring network, messages are transmitted in **packets** or **frames**, each of which contains a **destination address**. As the packet circulates past a repeater, the address field is read and if this corresponds to that previously allocated to the attached node, then the remainder of the packet is also copied. The frames are transferred sequentially, bit by bit, around the ring from one repeater to the next with each repeater serving to regenerate and retransmit each bit within the packet frame. Note that a **repeater** differs from a **transceiver** used in bus networks, since the latter does not automatically repeat and pass on data received at its input. A number of techniques have been devised for entering data into the ring from any end device connected to a repeater in such a way as to avoid contention and to make the maximum use of the data-carrying capacity of the ring topology. Two of the most common are the **slotted ring** which involves circulating a number of empty data frames around the ring and the **token ring** in which short, unique packets of data may be exchanged for much larger message frames. The operation of these techniques will be considered in Chapters 7 and 8.

A problem with ring systems that is not present with bus or star networks is the possibility of unwanted data or data fragments circulating around the network indefinitely, thus limiting the traffic-carrying capability of the system. With bus or tree topology, messages, once inserted into the media, propagate to the ends where they are absorbed by terminators so that the bus is clear of data shortly after transmission ceases. However, because the ring is a **closed loop** the data will circulate indefinitely unless specifically removed, since they will be regenerated at frequent intervals by the network nodes. There are various strategies used to determine *how* and *when* packets are added to and removed from the ring and these form part of a medium access protocol which is considered later.

6.2 POINT-TO-POINT DIGITAL COMMUNICATION

Many digital data communication processes are concerned only with simple point-to-point connection; a complex network with its routing, priorities and queueing systems is not involved.

A single, unshared line having no repeaters connects the transmitting and receiving ends between which each takes turn to exchange data. An example is the terminal/computer connection where a number of terminals are to be supported simultaneously, when the system is referred to as a **terminal-based network**. Generally, the terminals themselves have limited intelligence and play only a small role in control over the transmission process. Instead the computer to which they are connected arranges the reception and processing of characters conveyed from the terminal keyboard and display screen, controlling the exchange of a single character stream between the terminal and the computer. More sophisticated terminals handle a string of characters, displayed as a complete screen of information. Work stations and remote job entry terminals often possess quite considerable processing power (they are generally personal computers adapted for this purpose). This enables the volume of data and commands between terminal and computer to be reduced as the character manipulation and formatting may be carried out locally.

Whilst their control characteristics may vary, the exchange of information between computers and terminals, either as single characters or blocks of data, is carried out to certain agreed international data exchange codes. The most widely used are the American Standards Communication for Information Interchange (ASCII) and the CCITT International Alphabet No 5 (IA5). A third code is the IBM standard, Extended Binary Coded Decimal Interchange Code (EBCDIC). Some of the characteristics of these codes are given in Table 6.1.

Table 6.1

Data exchange codes

	ASCII	IA5	EBCDIC
Reference authority	USA	CCITT	IBM
No of bits	7	7	8
No of characters	128	128	256
Special features	parity bit	32 control codes 96 data codes	contains many unattached bit codes

Terminals used as point-to-point end devices employ either asynchronous or synchronous transmission, dicussed earlier in section 4.3. With asynchronous transmission, each character is transmitted independently with its own appended start and stop bit so that it is a fairly simple matter to control transmission from the sending end. Synchronous transmission is more complex, requiring either the transmission of an accompanying clock pulse signal or arrangements made to extract a clock signal from the data transmitted. Synchronous transmission can be faster and more efficient but requires greater control to be exercised locally by the terminal and hence is limited to the more sophisticated (i.e. intelligent terminals).

6.2.1 Polling

Several stations may be associated with a point-to-point link when it is referred to as a shared or **multipoint** link. This connects several nodes together but allows transmission between only one pair of nodes at a time. Often one of the stations assumes the role of a **primary** station with the remaining stations called **secondary** stations. Contact between the primary and the secondary stations is referred to as **polling** with the contact message transmitted called a **poll**. Each node is given an address which enables a secondary station to recognize messages sent to it from the primary station. Exchange between a secondary station to the primary station is permitted only when the primary station allows this. The primary station sends a poll message to the secondary station which may respond negatively, indicating no message for transmission, or may then send its stored message.

Two types of polling algorithm are commonly used: serial and hub polling. **Serial polling** takes place when the primary station sends a poll containing the address of each secondary station to each station in turn. **Hub polling** is used where secondary stations are located at long distances from the primary station and involves sending a common poll to all secondary stations. The poll is received by the first secondary station. If this has no data to send, it sends a poll to the next station. The process is repeated until all the stations on the link have been polled. If a station *has* data to send it proceeds to send these to the primary station. After receiving the data the primary station resumes polling from this secondary station since it will have already polled stations prior to the secondary station which has just completed transmission.

A major problem with polling is the overhead required in processing polls for which a secondary station has no data to transmit. Serial polling requires considerably more processing in the primary station than hub polling and in this sense is less efficient. If selective polling is required, then this is simple to accomplish with serial polling but requires additional logic where hub polling is used.

6.2.2 Echo checking

Terminals connecting to a host in a point-to-point system generally rely on **echo checking** to determine whether the receiving DTE has correctly received a transmitted element. With each key depression the character transmitted by the terminal is received at the remote computer and immediately retransmitted or 'echoed' back to the terminal. Only at this time is the character displayed on the screen, having traversed both directions in the duplex loop. Should the character displayed not match the character selected at the keyboard, the operator would be aware that a transmission error has occurred and would be able to transmit an agreed control character, e.g. a delete, to inform the computer to ignore the previously transmitted character.

The technique of echo error checking is inefficient in transmission bandwidth since each character is transmitted at least twice. This is not generally of importance in a slow terminal environment but in a data communication situation where a user is not involved and the process of error checking needs to be automatic then an **auto repeat request** is initiated using ACK and NAK acknowledgements as discussed in section 5.7.

6.2.3 Computer port allocation

Computers are limited in the number of external connections, called **ports**, that they may have, and it is wasteful to allocate permanently a port to a single terminal where it is likely to be inactive most of the time. Instead, methods of sharing ports between many terminals are applied. One technique used with those wide area networks making use of the PSTN, is to permit the terminals to compete for a port connection through the normal dialling process ('dial-in' modems). Where a small number of terminals are located in close proximity with each other (e.g. in the same room) a **cluster controller** may be used. This is a small computer port-allocation device requiring some limited cooperation with each terminal to permit a number of them to share a common line. It does, however, restrict communication to one terminal at a time. The cluster controller can provide a number of supervisory facilities, including buffering of incoming data to allow for diverse incoming data rate, error checking and automatic message retransmission. Format conversion may also be included.

The addition of substantial intelligence to a distant terminal has enabled port selection to be carried out on a dynamic basis. Two methods are in use. The port connection may be controlled by the system operator so as to connect one terminal input port to a given computer input port in accordance with a pre-arranged algorithm, set up at installation, and which can be altered dynamically in accordance with traffic conditions. Alternatively, a port may be associated with a destination address register and may accept data only from the destination address contained in this register. Port connection is then arranged by transferring destination addresses from one port to another. Data are transmitted carrying this address information and arrangements are included to transmit appropriate control signals (busy, continue or repeat) to deal with contention between terminals competing for the same port.

6.3 SWITCHED EXCHANGES

6.3.1 Terminal-switched exchanges

For large systems having many terminals this individual control process will be too cumbersome and a **terminal-switched exchange** (TSE) is used[1]. This is a device having a given number of asynchronous terminal ports (several hundred) and a smaller number of ports to which the computer(s) may be connected.

The operation is similar to a dial-up system with the operator first establishing connection by entering an address from the keyboard prior to transmission of the message. Assuming a port is free, the TSE will establish a logical connection between the terminal and the computer port. This is followed by the usual logging-in procedure by the operator and transfer of data. The operation of the TSE is thus made transparent to the user. Clearing of the connection is made through a specific control character entered by the terminal user which acts to free the computer port to which connection is currently being made.

6.3.2 The computerized branch exchange

A special type of TSE is the **computerized branch exchange** (CBX). This is a development of the private automatic branch exchange (PABX) which provides a company or organization with its own private telephone exchange linked by one or more wideband trunk connections to the PSTN.

In the CBX, digital switching is applied and the system is capable of handling digital speech transmissions, data work stations and, more recently, facsimile and other services (see Chapter 9). Dedicated port assignments are used for all attached telephones and devices. The CBX can be considered as a form of local area network having simplified routing arrangements. It uses PCM digitized speech signals which are integrated into common paths throughout the organization by means of TDM. Common-channel signalling is used, thus giving more effective use of the available multiple transmission channels.

Unlike the TSE, the CBX is able to provide the control signalling required for telephone connection (e.g. interconnection, ringing, speech interchange and stand-down) as well as supervisory functions such as call timing and charging. The data switching requirements are similar to the TSE, including addressing capability to accept differing transmission rates and calling/busy advisory signals. Direct connection to external data devices permits operation without a modem. The modem conversion to the PSTN if needed, is made on the basis of a high-speed modem generally incorporating TDM within the CBX. Digital speech signals are conveyed at 64 kbps for each speech channel so that, in addition to high-speed modem connection to the PSTN, there will be the need for codec conversion as part of the interface arrangements (see section 2.4.2).

CBX system architecture takes the form shown in Fig. 6.4. As a minimum requirement, the CBX will contain a data switching system, a call control unit, TDM facilities and high-speed group modem conection to the PSTN. An interface to multiple input lines operates as a synchronous TDM. Each incoming line is sampled at a given rate and the data buffered and organized

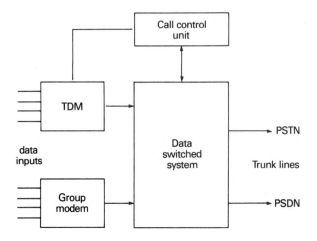

Fig. 6.4 CBX system architecture.

into frames. These are sent out by the switch at a high data rate (up to 140 Mbps) demanding a wide bandwidth trunk connection to the PSTN or PSDN.

For the larger CBXs a distributed architecture may be employed in which the central switch controls a number of subswitches interconnected through wideband coaxial cable or fibre optic links.

6.3.3 The public switched data network

Simultaneous transmission of multiple speech channels via the FDM mechanism of the PSTN was described in Chapter 2. Whilst the system can (and is) used for the transmission of digital information through the use of modems, it is not the ideal way to make use of the media for digital transmission. Further, the recent advances in PCM and digital technology make the transmission of multiple speech signals in digital form a more economic proposition. It also allows digital storage with store-and-forward operation for speech **datagrams**. For these reasons, increasing use is being made of PCM to convert multiple speech signals into an interleaved stream of digits through sampling at a rate of 8000 times per second and coding each sample to a level of 7 or 8 digital bits. The way in which the multiple-channel digitized signals are combined to form a single transmission system is described with reference to the **public switched data network** (PSDN).

In the UK, PCM transmission in the PSDN is utilized in a number of different ways. Transmission of a single data stream is offered with the **Kilostream** service which covers the same transmission speed range as the earlier FDM modem services (DATEL) at a rate of 2.4–64 kbps — adequate for a single digitized speech channel or low-speed data. This system makes use of existing four-wire circuits from the local exchange to the users' premises. A multiple data transmission service, **Megastream**, offers two standard transmission speeds. These are 2 Mbps, used mainly for connecting CBXs and with the ability to multiplex 30 digital speech channels, and a higher-speed transmission of 8, 34 and 140 Mbps for large-volume data transmission requirements. These services conform to the CCITT TDM hierarchy of transmission speeds shown in Table 6.2. A third service, applicable to long-haul WAN transmission, is the satellite relay scheme, **Satstream**. The speed range provided by this system is intermediate between Kilostream and Megastream, namely 41–1920 kbps. Finally, a private store-and-forward message switching system is in operation called **Primex**. This replaces an earlier telegraph system and provides up to 128 connections at 50, 75, 100 and 300 bauds plus a medium-speed facility operating at 1200, 2400 and 4800 bauds, all carrying full duplex connection.

Table 6.2

CCITT PCM hierarchy

Level	No 64 kbps channels	Approx. binary data rate (M bps)
1	30	2
2	120	8
3	480	34
4	1920	140

Two other services available for data transmission are the **Packet-switched Service** (PSS), with its links to the International Packet-switched Service (IPSS), and the Integrated Services Digital Network (ISDN).

6.4 PACKET SWITCHED SERVICES

PSS and IPSS play a very important role in the public data transmission domain. The principles of packet switching were considered in Chapter 5. These are applied to the PSS through an internationally agreed set of CCITT recommendations — the X series protocols, considered in Chapter 8, and implemented in the UK as the **System X** service.

6.4.1 System X

System X is a family of digital switched telephone exchanges developed for British Telecom, which are gradually replacing the analogue PSTN described briefly in Chapter 2. The characteristic features of a network using System X are:

a integrated digital switching and transmission,
b computer control using a stored program, and
c common-channel signalling.

These are implemented through a hierarchy of digital exchanges, comprising local exchanges based on a miniprocessor and a number of larger traffic capacity (trunk) exchanges incorporating a bigger processor. At the high end of the exchange size range the call throughput can be considerable, with 500000 busy-hour call attempts not being unusual. This requires not only a high processor operating speed but also that the processor does not execute too great a number of processor instructions per call. In System X many of the more routine but time consuming operations are devolved to dispersed microprocessors located in specific subsystems. Apart from these, other subsystems are required for high-density data exchange, international gateways and other tasks[2]. Data transmission between exchanges makes use of internationally agreed CCITT recomendations, the X series protocols (hence the name 'System X'), principally X25, X21 and X75 which are discussed in a later chapter.

Pulse code modulation (PCM) is used to convey digital information between exchanges using a 'slotted frame carrier' to CCITT recommendation G732. This provides the equivalent of two-way communication at 2.048 Mbps, structured into 32 channels of 64 kbps. Each channel is carried in sequence —1 byte from each in turn. The channels are numbered sequentially 0–31, with channel 0 carrying a code which identifies it as a 'start of frame' channel. Channel 16 is reserved for common-channel signalling and is often referred to as the 'D channel'. (We shall meet this again in connection with ISDN, where recommendation G732 serves as a basis for operations governing convergence of digital services.)

A key role in the network is the local exchange, and since this illustrates the characteristic features of the system as a whole its architecture and operation will be considered first.

6.4.2 System X local exchange

This is based on two major elements: a digital switch which interconnects a number of 64kbps data channels through a 2.048 Mbps multiplexer, and a combination of time and space switching controlled by a central processor (computer).

The structure of the time–space–time digital switch is shown in Fig. 6.5. Each 2.048 Mbps PCM system connects to a digital line termination unit (DLT) which fault checks and accepts 32 receiving and 32 transmitting

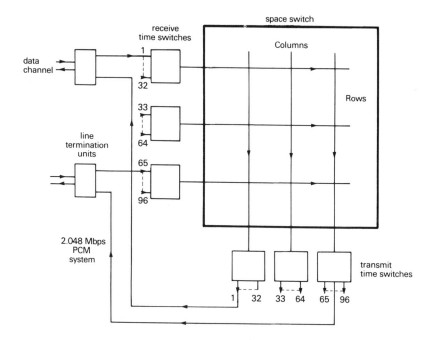

Fig. 6.5   A time–space–time digital switch.

channel signals. These are applied to time switches which convert the parallel 32 channel signals into two serial 1024 time slot highways, one for the receive channels (RTS) and one for the transmit channels (TTS). Other incoming 2.048 Mbps PCM system DLTs are connected to other 1024 time slot highways in the space switch as shown. The function of the time switch involves receiving a speech sample (8 bits) in its speech store, introducing a delay and then forwarding it in an alternative time slot. A control store maintains a matching set of locations for each time slot which holds the address of the speech store to be accessed at that time.

In operation, a given input channel is identified and a received word is written into the speech store in the RTS. At one selected time slot out of the 1024 this is read out and passed, via the connected space switch row, and selected by the addressed column for reading into a second speech store in the TTS. This can then be read out and transmitted via the appropriate DLT to the trunk PCM system. The speed of operation is at 8.192 Mbps, giving a time-slot allocation at every 125 ms. A 9-bit word is actually transmitted to include an additional bit for parity checking.

Software controlling the central processor consists of three elements: the operating system, the application code and the application data. The operating system, in addition to general supervisory tasks, also deals with such functions as storage allocation, job scheduling, overhead control, fault detection and recovery mechanisms. The application code determines the telephony function and user facilities supplied by the exchange, including new features not provided under the earlier PSTN system, such as three-party service, call waiting and barring, repeat call etc. The application data contain exchange-dependent data such as routing tables and directory number translation.

6.4.3 System X signalling system

The common-channel signalling (CCS) used in System X differs from the channel-associated signalling used in the PSTN. CCS is much more efficient in terms of processing operations per call and enables a wider range of control signals to be conveyed inaudibly to the user. The system also permits the signalling channels to be used for needs other than call control, e.g. network management and maintenance information.

The structure of CCS used in System X is such as to separate the two functions involved in the transfer of signalling information between exchanges and the call control process. Thus the interface between the message transfer control and the user system allows a number of separate pieces of information to be transmitted and acted upon. This is illustrated in Fig. 6.6 which shows the various categories of transferred information. The CCS uses one channel of the 2.048 Mbps PCM system with another channel reserved for synchronization, alarms, etc. so that, as stated previously, the PCM system actually provides only 30 message channels out of the 32 available[3].

6.5 VALUE ADDED NETWORKS

A carrier communication network, such as the PSDN, can be enhanced considerably in terms of user facilities by the addition of computer control of data transmission through switching processors and host computer interfaces. These enhanced user networks are known as **value added networks** (VANs). Their purpose is to provide a number of user services in addition to the primary service of providing a connection to the called party.

The PSS service described earlier is a VAN provided by British Telecom. A table of similar packet-switched public services throughout the world is given in Table 6.3. Together these form an International Packet-switched Service (IPSS) capable of exchanging information between cooperating nations through common X25 and X75 protocols. It is considered by British Telecom that the majority of their revenue in future years will derive from the VAN services on offer rather than from telephone calls.

Other VANs have been developed as commercial undertakings to provide

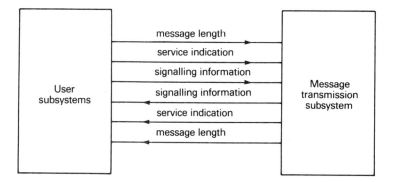

Fig. 6.6   Common-channel signalling.

Table 6.3

Packet-switched services world-wide	
Australia	AUSTPAC
Canada	DATAPAC
France	TRANSPAC
Germany	DATEX-P
Italy	ITAPAC
Japan	DDXP
Malaysia	MAYPAC
Netherlands	DATANET–1
New Zealand	PACNET
Norway	DATAPAC
Singapore	TELEPAC
South Africa	SAPONET
Switzerland	TELEPAC
USA	TELENET
	TYMNET
	COMPUSERVE

specific services such as videotext services, electronic mailbox to store messages, Graphnet for facsimile transmission, Telenet and Tymnet for computer host-to-host connection and Arpanet and JANET to provide facilities for the research and academic communities.

The advantages of a VAN from the user's point-of-view are:

a Only a small capital investment is required to gain access to the network.

b The organization providing the VAN takes full responsibility for its development and maintenance.

c Tariff charges are for the actual conection time to the network.

An additional advantage for most VANs is that connection is to the nearest interface processor and this can mean a local call connection in most cases.

Routing and control of the multiplexed packet stream lies with the vendor, not with the user. The VAN is, however, able to provide the user with a choice of processing facilities which may be evoked automatically or in response to user request, through additional control bits appended to the message transmitted. Some of these are:

a speed conversion,

b code conversion,

c delayed delivery, and

d priority transmission.

The two modes of packet transmission used in VANs are the virtual call and the datagram described in section 5.3.

Message routing has considerable importance for certain types of user messages. Several alternatives are applied in different VANs. A common technique is static routing in which packets between two nodes always take the same route, with a fixed alternative if the specified route is not available. Dynamic routing takes the traffic loading into account. Each packet is routed separately so that it is possible for components of a single message to take a number of different paths through the network.

In the following sections, some examples of VANs will be discussed, commencing with the earliest VAN, the Arpanet, which still provides a service to a widely scattered research community.

6.5.1 Arpanet

The first VAN to be implemented was the network of the **Advanced Research Project Agency** (ARPA) which commenced operation in 1967[4]. The network currently services about 100 sites located in university and research organizations within the USA and with links to Norway and the UK. The wide geographical distribution of the network is shown in Fig. 6.7.

Arpanet is a packet-switched network using public leased lines operating at about 50 kbps. Control is distributed throughout the network using two types of processors called **interface message processors** (IMPs) and **terminal interface processors** (TIPs)[5]. An IMP connects one or more host computers to the network, whilst a TIP provides end connection for up to 63 devices. Management functions are also distributed, with certain sites having specific responsibility for dynamic system modelling and traffic data analysis.

Arpanet operates on a datagram principle involving the transmission of separate messages. These may be up to 8063 bits in length. However, the

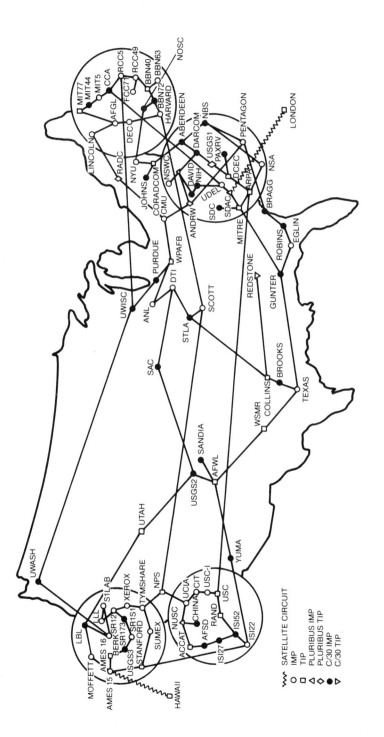

Fig. 6.7 Arpanet.

circulating packet contains 1008 data bits plus overheads so that the message is broken up at the IMP into eight separate packets, each of which needs to carry the identifier for the message as well as the packet number. In this respect, Arpanet differs from nearly all other packet systems. The facility for splitting a message into packets and reassembly of these to form the message after the packets leave the network, is generally expected to be carried out by packet assembler/disassembler units (PADs) external to the network system (see section 8.2.3), but in Arpanet this forms an integral function of the network. Further, Arpanet allows multiple routes to be taken between source and destination and there is no attempt by the intermediate IMPs to maintain a sequential packet order. This led to a particular problem with the network in which a limited set of buffers reserved for message reassembly at the IMPs could become filled with partly reassembled messages whose missing packets were held up in neighbouring nodes, unable to enter the destination node. This condition, called **reassembly lock-up**, is discussed in detail elsewhere[6]. It resulted in the development of fairly complex protocols to allocate the buffers separately for each message transmitted by the source node.

ARPA pioneered the use of layered protocols, that is sets of data handling rules, each set dependent on an underlying set referred to as the lower layer, long before international acceptance of such protocols was considered. The lowest of these layers was concerned with reliable communication between two IMPs. This provided transmission error detection and correction, routing and flow control for preventing congestion. An interesting feature of the IMP–IMP protocol is that it does not make use of negative acknowledgements (NAKs). If a packet is corrupted during transmission, it is ignored and a time-out mechanism initiates an automatic retransmission of the affected packet. The next layer is an IMP-to-host protocol providing for exchanges between host computers to establish and maintain communication between processes and user jobs on the remote computers. A virtual terminal protocol enables direct connection to be made for multi-terminal input via TIPs to the host computer. Several higher-level protocols such as file transfer, electronic mail and remote job entry foreshadow later development of the ISO upper-layer protocols now widely accepted in network design. Another 'first' was the use of adaptive routing, initially concerned with estimating the minimum expected delay for routes out of a given node and forwarding this information to neighbouring nodes in the network. It is carried out by the IMP which can apply the information to determine a path having the least delay for a given message. Such a path will vary with time and is modified by update information circulated by the IMP as described in section 5.6.1.

The inclusion of such complex multi-frame transmission protocols, adaptive routing and VAN services such as mailbox and traffic monitoring has resulted in a packet format that is necessarily complex. This is because it needs to carry not only the source and destination addresses, message identity, packet sequence number and framing control information, but also a series of check characters for error control, acknowledgement and special instructions such as priority, discard, trace etc. Thus, in terms of transmission efficiency, very little actual data are carried in each frame compared with later networks. These latter operate in a virtual circuit environment rather than the message/datagram system applied in Arpanet and require much shorter header information. As a result, the bandwidth demands are smaller and a more economic operation is obtained.

The operational facilities of Arpanet are, however, immense and the concept has been extremely valuable as a testing ground for packet-switching techniques in general. One such technique is that of TRACE in which a trace message may be inserted into the packet format by a given host. This permits subsequent analysis to be carried out to identify packet arrival times, queueing times, route taken etc., enabling information to be built up on how the network responds to a particular traffic and configuration situation.

Arpanet is probably the most thoroughly researched network yet developed and has provided the basis for all subsequent packet-switched VANs such as PSS and its derivatives, Cyclades, Datapac, Tymnet, Telenet and many others.

6.5.2 Janet

The Joint Academic Network (JANET) includes a VAN that serves the academic community in the UK.

JANET is a packet-switched network operating with leased lines to PSS operating standards, and links 52 university sites in the UK, government research council establishments and a number of polytechnics[7]. A feature of the network is the ease with which it may be linked to LANs and host computers at individual node sites. The protocols in use conform broadly to the OSI seven-layer model but there are differences in implementation in order to accommodate some of the higher-layer facilities required in advance of agreed international standards becoming available.

The elements of JANET are:

a   A packet-switched service conforming to levels 1–3 of X25 protocol format.

b   JANET packet-switched exchanges (JPSE).

c   PADs to link to user terminals.

d   'Rainbow Book' ISO protocols.

e   Use of British Telecom leased links as the basic carrier.

f   Trunk lines between exchanges.

'Rainbow book' protocols refer to a series of OSI–related protocols developed by or for the universities and identified by a specific colour. Their function and relationship with the ISO level specification is shown in Table 6.4. A brief discussion on these will be found in Chapter 8.

Table 6.4

JANET communication protocols

Title	Function	Colour
NITS	Network-independent transport network service over X25	Yellow
NIFTP	Network-independent file transfer protocol	Blue
JTMP	Job transfer and manipulation protocol	Red
JNT	Electronic mail transfer based on Arpanet format and using blue book for message transfer	Grey
TS29	Character terminal protocol over transport network service using CCITT recommendations X3, X28 and X29	Green
CSMA/CD	Protocol and interface over a bus network	Pink
CR82	Protocol and interface over Cambridge ring network	Orange
Interactive	A protocol providing efficient operation of full-screen interactive terminals	Fawn

These protocols differ functionally from the OSI reference model which has seven layers whereas the model developed for JANET has only three. These are:

a **The application protocols**. These protocols are not layered in a strict way, although one protocol may exploit another to perform a common function.

b **A convergence protocol**. This protocol establishes a technology-dependent independent service upon which the application rests. This boundary is called a transport service.

c **Subnetwork access protocols**. These are the protocols necessary for the operation of the particular network technology in use. For example, one protocol may be used for a ring technology and another for a bus technology.

In Fig. 6.8 the functional parallels between these layers and OSI are shown. The lowest protocol layers are compatible with the British Telecom PSS

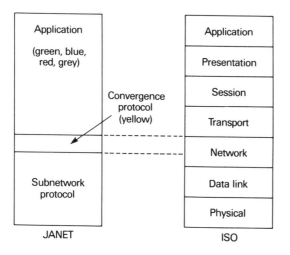

Fig. 6.8 JANET 'rainbow' protocols.

service, which provides an internationally agreed series of X-series operations protocols. The upper two layers differ in some important respects to current OSI standards. It is the intention to migrate eventually from the present 'rainbow book' protocols to the full ISO OSI standards. This, however, involves the identification of adequate and agreed interpretations of the standards and this agreement is not yet a reality. The new products must co-exist with existing products before they can displace them and the entire process is expected to take several years to accomplish.

A number of value added services form part of the network. These are:

a name server;

b gateway services;

c network status and help facility;

d store and forward for mail and job submission; and

e software directory diagnostics facility.

The architectural form of the network consists of ten **JANET packet-switched exchanges** (JPSEs), each located at a **Network Operations Centre** (NOC). The exchanges are connected by trunk lines operating at 9.6 kbps and 48 kbps. The higher-speed links are based on the British Telecom Kilostream service. Both use X21bis protocol at level 2 (see section 4.5.4). Each NOC contains a switch based on a GEC4000 series communications processor —this is the JPSE. The largest of these processors has presently (1985) a switching capacity of about 750 packets per second which is adequate for the traffic levels experienced with the system[8]. Software for the switch is contained in an external memory (disc) together with other software require for switch maintenance, fault diagnosis and other support functions.

Each NOC has at least two lines to neighbouring NOCs which provide a measure of alternative routing. The network is a decentralized one with control residing at each NOC, where services are provided to about 120 other sites. The principle is for each site to be connected to its nearest NOC (geographically) so as to form a series of star-connected networks. Each of the NOCs is connected via JANET to a Network Control Centre (NCC), which houses the Network Management Unit (NMU). The NMU collects the periodic reports from the NOCs, and on the basis of these is able to review total network operations and, in some cases, to detect certain classes of fault. The network can continue to run in the event of the NMU failing, though clearly monitoring will cease.

Five of the sites, Rutherford Laboratory, Manchester, London and Edinburgh University Regional Computing Centres and the South-west University Computing Centre at Bath are connected by 48 kbps leased lines and operate under the British Telecom Kilostream public service. The remaining sites presently operate under 9.6 kbps leased lines, although there are plans to upgrade these to 48 or 64 kpbs with trunk lines transferred to the British Telecom Megastream service operating at 2.048 Mbps[9]. A feature of the JANET network is that the end-site connection can be to a large number of user terminals or, alternatively, to a local area network. In the former case a series of packet assembler/disassembler units (PADs) are employed, each containing multi-input concentrator facilities. The LAN connection makes use of a subnetwork access protocol suitable for a 10 Mbps slotted ring or to CSMA/CD token ring/bus technology conforming to the IEEE 802 standards (see Chapter 8).

JANET is also connected through designated end sites to other VAN networks, specifically the Arpanet and the European Academic Network (EARN). These linkages are provided through protocol converters (gateways) located at the end sites.

SUMMARY
Chapter 6

Communication networks make use of two basic ways of conveying digital data. The first is to **switch** the data from node to node across the network and the second is to **broadcast** the data to all the nodes in the network, relying on the carried address information to determine the ultimate destination.

The first method is applicable to the wide area network, discussed in this chapter, and the second to the local area network to be considered next.

The way in which data are transferred or broadcast is

dependent on the way in which the nodes are inter-connected — the topology of the network. Several different forms are available, mesh, bus, tree and ring. Point-to-point communication is the basis for more complex network interconnection and its operation can be studied with reference to a simple computer–terminal linkage. Control for such a connection can reside with the computer, or be shared between the computer and a terminal equipped with appropriate control logic. The problem of connection for a number of terminals to a single computer is overcome by the use of **cluster controllers** or **terminal-switched exchanges**. A special form of switching exchange is the **computerized branch exchange** which is a digital equivalent of the private telephone exchange.

A network representing the digital equivalent of the public telephone network is the **public switched data network** (PSDN). A number of different digital communication services are available over this network, each having a different range of data speeds and service facilities. One important PSDN service is a packet-switched service (PSS) which is available in very many countries to similar CCITT data interchange recommendations.

When the public carrier systems are augmented with computer control this enhances the services available to the user and they are referred to as **value added networks** (VANs). Very many of these VANs are in operation, providing different services to different user communities. Two of these, which are particularly valuable to the academic and research communities, are Arpanet in the US and JANET in the UK.

**PROBLEMS
Chapter 6**

P.6.1 Describe the basic differences between a wide area network and a local area network in terms of:

a structure,

b operation.

P6.2 The techniques of passing information from node to node across a broadcast network differ according to the type of configuration employed.

Compare the methods used for bus and ring networks.

P6.3 What is a value added network? Give three reasons why a corporate user would wish to make use of such a network.

P.6.4 A poll and its response takes eight characters. Assuming that a network connection operates in half-duplex mode, that the data rate is 9600 bps and that the modem used requires 10 ms to operate:

a Calculate the number of polls per second if there are no data to send.

b Given eight secondary stations, what is the average delay to send a data message of 100 characters from the last secondary station to the primary station?

(One character = 8 bits.)

P6.5 **a** Explain why a terminal-switched exchange cannot be used as a substitute for a computerized ranch exchange.

b How is the rate of 64 kbps arrived at for single speech channel transmission?

P6.6 A store-and-forward network has the ability to accept data at one transmission speed and transmit it at another speed. What are the limitations of this process in a practical network?

7

LOCAL AREA NETWORKS

We can define a **local area network** (LAN) as a 'communications network that provides interconnection of a variety of data communication devices within a small geographical area'[1]. Other characteristics follow from this definition. Because the network is confined to a small area it is not necessary to make use of public networking facilities, with their undesirable overheads in terms of transmission via an undefined path through an undefined number of switching nodes and the use of complex routing algorithms. With the freedom to lay cables and design the network specifically for one group of users, data transfer becomes simpler and considerably higher transmission speeds can be employed. As a consequence, simple stop-and-wait protocols are adequate and the transmitted data packet can contain more information (particularly related to addressing) since a wider bandwidth is available.

With small transmission distances, the media contribute little distortion to the signal and fewer switching nodes are encountered. This results in error rates for the LAN of the order of 1 in 10^9, which is several orders of magnitude better than the WAN. One further simplification arising from the limited geographical area is that ownership of the LAN can be by a single organization, so that multi-layer protocols providing for a wide range of possible user requirements are not needed and the whole design approach can be less cumbersome.

Local area networks are required by a number of specific user groups. The most common of these is where it becomes necessary to link together a number of personal computers to connect them to a remote host computer through a common channel or in order to share resources, e.g. memory, data bases or specialized applications. The development of public electronic mail systems and the newer requirements, such as machine tool control and the exchange of computer graphics information, will accelerate the application of LANs in the office and industrial environment. In contrast, at its simplest level, a LAN may be an economical way to link a small number of computers (perhaps only one) to a number of remote terminals, printers or other digital devices.

Two rather specialized LANs are also in use. One is the high-speed local network (HSLN), designed primarily for interconnection of a limited number of high-speed devices such as mainframe computers in a large data processing site. A second, which was discussed in the previous chapter, is the computerized branch exchange (CBX). This differs from most LANs in that networking operations are controlled from a single central node. It is less flexible than the networks discussed in this chapter and since it is designed to handle both speech and data connections it is often considered as a digital extension of the telephone exchange.

7.1 LAN TECHNOLOGIES

Two broad divisions in the topology used for LANs can be distinguished: **linear systems**, typified by a bus or tree topology and **ring systems**. The operating constraints on the two systems are quite different. In the linear system the network connecting a number of communicating nodes is a passive device and simply accepts any information presented to it for transmission in both directions along the media to all the stations on the network. Thus the individual stations must **contend** for use of the network through their communcating nodes. A prime consideration in linear systems is how to design a suitable **contention protocol** to avoid the situation where more than one station tries to make use of the network at the same time.

A different kind of problem exists with ring networks. Here the nodes play a more active role, in which they accept information from one of their neighbouring nodes and retransmit this to the next node in the chain (only one direction of transmission around the ring is permissible). It becomes rather simpler to ensure that only one station can make use of the network at a given time, but the endless nature of the media connection does give rise to a difficulty. Unless active steps are taken to prevent it, a message, once introduced into a ring, will circulate indefinitely. Thus it is necessary to include **data termination procedures** into the ring protocol. This needs to cover not only the situation where the transmitting station fails to remove its own message, when it has circulated the ring once, but also the possibility that random digits, noise and fragments of message data may be present in the network and require removal.

The techniques of contention removal and data termination procedures thus dominate the technology of the two generic types of systems. Essentially, these are matters of protocol design formulated within the controlling software of the network. How complex they need to be, however, is determined by the physical arrangement of the network, the use to which it is put, and the length and data rate for the message being conveyed. In this chapter these physical considerations will be discussed, leading to the description of the range of alternative networks that are possible with current technology. The design of communicating protocols will be considered in the following chapter for both WAN and LAN systems.

7.2 SIMPLE HANDSHAKE SYSTEMS

The simplest form of LAN is one which makes use of the RS-232c parallel connection recommendations described in section 4.5.1 and which can connect two devices capable of operating in this way via a null modem (section 4.5.2). This is essentially a point-to-point linear network in which the possibility of contention has been excluded by a limitation to only two end stations connected to the network at any one time. A typical example would be a network connecting a number of personal computers to a shared printer. Only a few kinds of devices may be supported, and this type of LAN provides little more than the ability to transfer files and share high quality printers. Only one connection at a time is feasible. Other devices attempting to connect would receive a 'busy' response and try again later.

This type of network is, however, effective for its limited purpose and can be cheap to install and maintain. The controlling signals for the RS-232c connections and additional signals required to interrogate the user for his address, to indicate 'busy', 'clear' etc., need to be provided through a hardware/software control algorithm. Two methods have found wide usage. One is to make use of the operating system within the personal computer to carry out all command network functions under control of an added software program. The second is to set up a subnet of hardware nodes, each consisting of a small microprocessor, to which the RS-232c devices may be connected.

7.2.1 Software systems

Typical of the software systems are the networks marketed as LANLINK[1] and NETWARE.[2] The network control programs are contained on one or two floppy disks which are read into an IBM PC/XT compatible system operating under PC-DOS or MS-DOS. This extends the normal DOS operating commands such as COPY and DIR to include a number of commands for the control of files being copied or transferred and to provide user status information. Device recognition can be through software addressing (each device is identified during an initialization phase) or, in the case of non-intelligent devices, a hardware recognition unit is required. One

[1] LANLINK: Intercompany Communications Technology Ltd., London, UK.
[2] NETWARE: Novell Data Inc., USA.

advantage of this close association with the normal operating system is that proprietary files such as LOTUS 1-2-3, dBaseIII and Wordstar can be transferred easily by command across the network. A separate RS-232c serial port is required for each connected device. In operation, the system is limited to a small number of connected devices, typically 16 (star) or 32 (ring) and provides a data communication rate up to 500 kbps. NETWARE represents a major extension to this type of LAN software control. It operates effectively as a network file manager running under the PC operating system. NETWARE is capable of controlling a number of other proprietary LAN hardware systems including OMNINET, ARCNET and IBM token ring, and will operate with a large number of user application programs.

7.2.2 Subnode systems

The use of a small dedicated microprocessor to manage interconnections relieves the personal computer of this task and enables a more flexible system to be used. Two successful systems of this type are CLEARWAY[3] and INFAPLUG[4]. Both make use of an 8-bit control processor.

CLEARWAY consists of a number of hardware nodes, each containing a Z80 microprocessor into which the network devices are connected via the RS-232c serial connection. In effect, the network appears to the computer as a serial port. Connection of up to 99 devices are possible, each identified by a station number. Operation is achieved by running a terminal emulation program within the personal computer. This allows the keyboard to address commands to the node units in a similar manner to the software systems. The network uses a register-insertion method of sending data packets around a ring (discussed later) to which the Z80 nodes are connected. Each node buffers packets transmitted between logically isolated ring sections. Data speed around the ring is at 56 kbps, but for users' data this is reduced by at least 16% due to protocol overheads.

Commands to the nodes are given in ASCII characters and take the form of a dialogue between the user and the system to set a number of parameters, e.g. node number (1–99), line speed selection (50–9600 bps), flow control method, choice of parity, configuration control character and node title. It is essentially a dialogue to establish connection between two specific end nodes, although, of course, several such pair connections can share the line (the addressed node will indicate 'busy' if a line not available). Multiplexed host connection or a broadcast facility is not possible.

Extensive systems of this type are the MULTINET and SIMPLENET systems marketed by Nine Tiles.[5] These also use a register-insertion method

---

[3] CLEARWAY: RTD Ltd. Farnborough, UK.
[4] INFAPLUG: Infa Communication Ltd, UK.
[5] MULTINET: Nine Tiles Computer System Ltd, Cambridge, UK.

of packet transmission with a ring network. Each node on the network includes its own CPU which handles all the low-level protocols required for setting up and maintaining virtual circuits, partitioning the data into packets for transmission, and ensuring that each packet is acknowledged etc.

In its simplest form MULTINET can act as an RS-232c switch connecting printers and VDUs into minicomputers, and software is contained in the node ROM for this purpose. This not only reduces the load on the host computer but also insulates the network from the effect of interrupts occurring in the host computer and variability in the clock rate. The data rate with the system is fixed at 250 kbps with communication to the host(s) via 192 kbps RS-232c connection. MULTINET can also support direct connection to the host computer (e.g. on an IBM PC) and permits multi-ring nodes to be connected via a node gateway.

SIMPLENET adds a series of servers and diverter modules to the basic network system. These act to permit disc files and other resources contained in the host machine to be made available to other devices in the network and permits the distributed operation of many user application programs. The message structure is arranged to provide a means of inter-networking between a number of different operating systems thus allowing an 'open file' access across the network.

7.3 LINEAR NETWORK SYSTEMS

Full linear networks achieve their interconnection by means of a passive bus connection with nodes attached to it at intervals along its length. The limitation in numbers of nodes connected is dependent on the transmission rate and the waiting time that can be tolerated before a connection is made. There is also a limitation on the physical length of such a bus due to the attenuation characteristics of the media and the transmission rate used. A linear network designed to provide a reasonable service within these constraints is the **Ethernet System** developed initially by the Xerox company in the 1970s[2].

7.3.1 Ethernet

The original Ethernet was intended to provide communications between personal computers and to share facilities in an office environment. This needed to be carried out cheaply in terms of the ancillary controlling devices and the communications media used. It was also designed specifically for computer connection where the traffic takes the form of short and intermittent 'bursts' of data. Earlier methods involving polling or TDM were not so well suited to handling data and were expensive and complex to install.

The medium chosen for tranmission was coaxial cable, although other media can be used, e.g. twisted pair or fibre optic cable. The bus technology employed uses a broadcast method originally developed from the radio communication experiments carried out by the University of Hawaii with its ALOHA network[3]. This applied a contention method of sharing the bus and the complex collision avoidance protocol required is still the major drawback of the Ethernet system. An Ethernet system for LAN is shown in Fig. 7.1a.

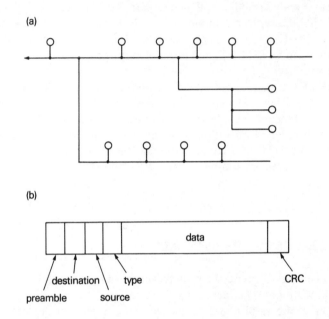

Fig. 7.1 Ethernet. (a) Bus/tree network structure; (b) frame format.

There is no central or master control point. The network can have branches like a tree structure and new taps and branches can be added easily. The taps to the cable are connected to **transceivers** acting as nodes at each station. The transceivers receive signals passing through the cable and are able to transmit signals sufficiently strongly to reach the far points of the cable so that regeneration of the signal at the nodes is unnecessary.

Ethernet comprises two main units: a transceiver and encode/decode unit which also carries out the detection of transmission and protocol operation in the event of a collision; and a controller and data encapsulation unit which carries out the recognition of packets for a given station, the detection (but not correction) of errors and managing the link connection.

A major cost of the network lies in the provision of transceivers, since the actual network itself is simply the cable and its tapping points. A main consideration in the design of transceivers is that if a fault develops in the unit then it must be automatically disconnected so as not to affect signals passing along the bus. Otherwise, the tasks required to be carried out by the node are relatively simple and this makes for cheapness in the overall design.

Ethernet is a packet broadcast transmission system. When a transceiver transmits a packet it is received by every other transceiver; the packet carries its destination address and only the station to which it is addressed may accept the packet. Other stations ignore the packet as it goes by along the cable. Control of packet transmission is thus distributed throughout the network; each station must make its own decision as to when to transmit. The problem is that two stations might attempt to transmit at the same time; their packets would then interfere with one another, i.e. 'collide', and cannot be received correctly. When two stations do accidentally transmit at the same time, they must both be able to detect the collision and to retransmit, at different times, new copies of the packets damaged by the collision. This means that 100% channel utilization can never be achieved. Ethernet can, however, come closer to this limit where polling and other techniques cannot, since the overhead of allotting time to stations that have no information to transmit is not present.

An Ethernet station transmits variable-length packets at a speed of millions of bits per second, typically 10 Mbps. The transceiver listens to its own transmission. If it notices a difference between what it is transmitting and what it is receiving, it knows that a collision is taking place. The transmission is then abandoned and the controller reschedules it for retransmission at a later time. Usually there will not be a collision the second time. If there is, a third attempt is made and so on (up to 16 attempts are permissible in an Ethernet system). In this way control of the network is completely distributed and alternative routing mechanisms are not needed.

7.3.2 Network control

The controlling technique which arranges this listening, checking and retransmission forms part of the **protocol** for the network. Protocol design for LANs is considered in detail in Chapter 8. Here the technique is discussed in general terms.

The protocol applied in Ethernet (and now widely adopted for other networks) is known as **carrier sense multiple access with collision detection** (CSMA/CD). The essence of the technique is first to listen to the signals broadcast along the bus network and to seize the opportunity to transmit a message when no activity is detected. Since more than one station may detect no activity and hence send at the same instant, there must be a mechanism to detect a collision and deal with it.

The process may be summarized as a set of rules:

1 If the medium is quiescent, transmit the message (packets).

2   If the medium is busy, continue to listen until the channel is quiet; then transmit immediately.

3   If there is a collision, wait a random amount of time and repeat step **1**.

4   If a collision is detected during the transmission of a message, suspend transmission and transmit a brief jamming signal to let all stations on the network know that a collision has occurred.

5   After transmitting the jamming signal, wait a random amount of time and repeat step **1**.

Information is transmitted along the network in a series of data packets, referred to as **frames**. Information packets from an end device are conveyed to the data encapsulation unit which assembles the packets into frames for transmission and adds a preamble and frame checking sequence as shown in Fig. 7.1b. The Ethernet format also includes a **source** and **destination address**, both 48 bits long, and a field to indicate the start of the frame.

A **type field** is appended to the frame, applied by the user to specify the protocol relevant to the device attached to the node, and finally a **16-bit cyclic redundancy checksum** (CRC) of the address and data fields for the end-to-end detection of transmission errors is appended.

Correct operation of a contention LAN requires that all stations agree on whether a collision has actually occurred. To ensure this, Ethernet specifies a minimum length for data frames so that the transmitting station will still be in the process of transmitting a frame when a collision is detected. This minimum length includes a limited data field of 46 bytes and is 72 bytes in length. The maximum data field that can be included is 1500 bytes.

During data reception, the receiver detects when a frame begins by recognizing the presence of a carrier, i.e. a sequence of bits constituting the frames. It then uses information contained in the preamble to determine the bit phase. This is made easier by the Manchester type of coding used for bit transmission (Fig. 4.4d), which ensures that a transition from 0 to 1 or 1 to 0 occurs with every encoded bit. It is the regularity of these transitions which serves to permit extraction of a clock signal and thus achieve synchronization with the transmitted information.

The first action to be carried out by the station is to check the destination address carried by the packet. If this does not match its own address it ignores the packet. If it agrees, the receiver reads the complete packet and detects when the carrier ends. It checks that an integral number of 16-bit words have been received and that the checksum in the last word is correct for the data transmitted. If not, an error routine is entered.

A **controller** initiates the retransmission of the packets. This also recognizes the **mean waiting time** for the cable (equivalent to an end-to-end trip delay from transmitter to receiver and back again). If a second attempt at retransmission is needed, the controller doubles this mean time and doubles it again on the third attempt and so on. This is known as a **binary exponential back-off** algorithm. In this way it attempts to adjust its behaviour to the load conditions at a given moment so as to minimize the probablity of collision.

7.3.3 Collision behaviour

When the load on the network is low, collisions are rare and the mean delay time rarely exceeds the minimum value of one round trip delay. Also, if a station, having control of the network, sends a long packet it is likely to shut down the other stations and minimize the number of collisions. Thus the Ethernet system favours a **lightly loaded** situation and **long packets**. Some useful diagrams illustrating this are given by Stallings[4].

The detailed behaviour in the event of collision between two stations X and Y separated by an inter-station distance, D, is shown in Fig. 7.2. Assume that X initiates a message packet on the bus as shown in Fig. 7.2a. This travels towards station Y and, just before it arrives, we will assume Y also commences transmission (Fig. 7.2b). The two messages collide and detection of this result by Y (which is nearest to the point of collision) will immediately cause a cessation of Y's message and the transmission of a short jamming signal. This takes a finite time to traverse the distance D between the two stations, so that station X does not cease transmitting immediately and will not do so until the jamming signal arrives (Fig. 7.2c). From this behaviour we can deduce that the maximum amount of time required to detect a collision is just twice the propagation delay and that the time spent in collisions and their detection is dependent on the length of the message frame, being short where the frames are long relative to the propagation delay.

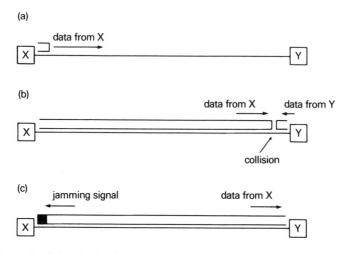

Fig. 7.2 Collision behaviour.

7.3.4 Ethernet summary

Ethernet with CSMA/CD protocol is now widely used in LANs and has been demonstrated to operate extremely well in a variety of circumstances. Simulation studies of highly loaded networks have shown that efficiencies of around 97% can be achieved. This is due, principally, to the matching of the system to the 'bursty' traffic characteristics of computer communication. It is less successful under heavy loading conditions and, additionally, the network response is statistical so that there is always a finite probability of an arbitrary long delay occurring before access is obtained. It is, on the whole, however, a successful system and forms the basis for very many commercial LAN systems and an IEEE networking standard discussed in the next chapter. To summarize, the major characteristics of Ethernet are:

a **Topology**: tree-shaped, composed of separate bus segments.

b **Medium**: Coaxial cable, 50 ohm impedance.

c **Signalling system**: baseband, bit-serial, Manchester encoded.

d **Data transmission rate**: 10 Mbps.

e **Maximum station separation**: 2.5 km.

f **Maximum cable segment length**: 500 m.

g **Maximum number of stations**: 1024.

h **Access method**: CSMA/CD.

i **Frame**: variable length (72–1526 bytes).

j **Addressing**: both source and destination address included, each of 48 bits in length.

Although the data transmission rate is 10 Mbps, the network is shared by every user on the system and since each frame contains control information as well as data, the actual useful data transfer rate between any two devices is much less than this. A rate of 1Mbps would be a realistic figure in an average network. The coaxial cable specified has 50 ohm impedance and is rather bulky. A less costly form of interconnection uses thinner 75 ohm cable, as used for domestic television. This form of Ethernet has been referred to as 'cheapernet' and operates with a reduced overall transmission rate.

7.3.5 HYPERchannel

An HSLN using similar techniques to that of Ethernet is the HYPERchannel network.[6] This operates a modified form of CSMA known as carrier sense multiple access with **collision avoidance** (CSMA/CA) and is designed to avoid increasing delays caused by packet collision. It finds its main use in the interconnection of main frame and specialized peripherals, such as mass storage devices, where transmission delays cannot be tolerated.

CSMA/CA carries out a listening operating before transmitting, as with CSMA/CD. If no other device is detected then data transmission can go ahead. If no collision occurs the destination device sends back an acknowledgement packet. This cannot itself be subject to collision since a different channel is used which is dedicated (for a defined period) to the receiving station. If a collision *does* occur then no acknowledgement is sent and the sender uses its own pre-allocated time slot to send a repeat message. The method is a combination of contention protocol and a time division multiplex system involving four separate channels each operating at 50 Mbps. These are pre-allocated to permit stations to transmit only within certain time slots without any possibility of collision occurring. If all these time slots happen to be unused at any particular time, they become available to all the linked devices. Should a collision occur, then the sender uses his pre-allocated time slot to repeat the data frame. Only a small number of connected devices are possible, with the connection range limited to the size of a large computer room.

7.3.6 Other linear networks

Several implementations of baseband bus networks have been made with the limited intention to link personal computers together in order to share a common printing device or to distribute functional operations. These use a simplified form of CSMA/CD in which the speed of operation is reduced and numbers of connection devices limited, to minimize the possibility of collision. Early designs of this type were CLUSTER/ONE,[7] originally designed to link Apple Model A computers together, and ECONET,[8] designed to link together a number of Acorn Computers (much used in primary education in the UK).

With CLUSTER/ONE the file server (which itself is an Apple computer) incorporates a hard disc unit which serves all the Apple PC's. When a linked computer is switched on, the file server automatically loads it with the operating system and allocates any file work areas requested by the user. In effect, each Apple work station then performs as would a normal Apple PC using normal disc drives. The print servers on the network (there can be more than one) allow various types of printer to be shared by all the PCs on the network. This network uses a form of CSMA/CD protocol based on

[6] HYPERchannel: Network System Corp, USA.
[7] CLUSTER/ONE: Nestar, Ltd, UK.
[8] ECONET: Acorn Computers Ltd, UK.

multicore cable (not coaxial) operating from the RS-232c port. The transmission rate is limited by the RS-232c connections to about 240 kbps and the number of devices to 64. More recent developments of CLUSTER/ONE apply a token-passing protocol and support up to 255 nodes at a data rate up to 4 Mbps.

ECONET operates very much as the CLUSTER/ONE network. The cable is four-wire twisted pair and up to 255 devices can share the same network. The data rate is 210 kbps and the maximum length of the bus is 1 km. A simplified form of CSMA/CD is used and a shared disc file is provided locally on the Acorn PC used as control.

7.4 RING SYSTEMS

A ring topology consists of a number of **repeaters** forming part of the ring station. Each repeater is connected to two others by unidirectional transmission links to form a single closed path as illustrated in Fig. 7.3. Data are transferred sequentially, bit by bit, around the ring from one repeater to the next with each repeater serving to regenerate and retransmit each bit within the packet frame. Note that a repeater differs from a transceiver used in bus networks since the latter does not automatically repeat and pass on data received at its input.

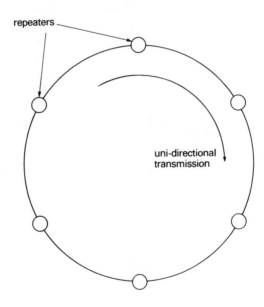

Fig. 7.3 A ring network.

Repeaters carry out three main functions:

a message insertion,

b message reception, and

c message removal.

Messages are transmitted in packets or frames, each of which contains a destination address. As the packet circulates past a repeater, the address field is read and, if corresponding to the station to which the repeater is attached, the remainder of the packet is also copied. The insertion/reception functions are thus carried out much like the nodes on a bus system, but **message removal** is quite different. For a bus or tree topology, signals inserted on the line propagate to the ends where they are absorbed by terminators so that the bus is clear of data shortly after transmission ceases. However, because the ring is a closed loop, the data will circulate indefinitely unless removed, since it is regenerated at frequent intervals around the ring. There are various stratagems which may be used to determine how and when packets are added to and removed from the ring. These form part of the protocol used in controlling the network, discussed later. Here we consider some alternative physical arrangements for a ring system.

7.4.1 The slotted ring

The **slotted ring** or **empty slot system** was originally suggested by Pierce in 1972[5]. One or more skeleton packets or **slots** circulate continuously round the ring. Their number is fixed and depends on the length of the slot together with the total length of the ring.

Typically, there will be very few slots on a ring. Consider, for example, a 100-station ring with an average spacing of $D=10$ m between stations and a data rate of $R=10$ Mbps. A typical propagation velocity for signals in a line medium is $P=2 \times 10^8$ m/s. Thus we can define the average 'bit length' of the link between two stations as the data rate times the inter-station spacing divided by the propagation velocity, i.e. RD/P. This gives an average value of bit length for the link of 0.5 bit. If we assume the delay at each repeater to be 1 bit, then the total bit length of the ring is just $1.5 \times 100 = 150$ bits, which is enough for four slots only in this example.

At ring start-up, one repeater or device generates one or more slots and sends these around the ring (initializing the ring format). If they return to the sender then this indicates that the ring must be complete and transmission of message information can commence.

A station with data to transmit partitions this into fixed-length frames, generally an integral number of 8-bit data bytes. It then waits until the repeater immediately preceding it in the ring passes an empty slot to it. The data packet is then appended to empty data fields in the slot as it passes through the repeater, after first setting the full/empty flag bit to full and placing the destination address in the header. When it reaches its destination device, the repeater reads the information into its own buffers without clearing the slot. This is later cleared when the slot (still flagged full) has circulated the ring completely and again reaches the sender. The sender recognizes the slot, either by its address content or by counting the number of slots circulating the ring, and sets the flag to 'empty' again so that the slot is free to be used by another sender. In one form of the system, the sending device can simply replace the slot with further data, overwriting the existing information, thus retaining the slot for a further period. If the destination device is unable to read the slot information, this fact can be flagged to the sender so giving a form of negative acknowledgement. Thus a station cannot transmit another frame until this slot returns. The slot also carries two response bits, which are set by the addressed station to indicate data accepted, station busy, or data rejected. The full slot makes a complete round trip, to be marked empty again by the source station as described above.

It is necessary to check that slots with data not accessed by the recipient after a given number of revolutions, are emptied and this is a job given to one particular device on the ring, known as the **monitor**. This monitoring facility is a feature of most ring systems.

7.4.2 The Cambridge ring

An example of slotted ring operation is the **Cambridge ring** devised by the University of Cambridge[6]. A station on the ring includes a repeater to restore pulse levels before circulation and an access box which transfers data recognized by the address contained in the packet format as that associated with the station. Hardware logic is included in the access box for error checking and various housekeeping and protocol logic procedures (considered later when we come to a discussion on LAN protocols).

The repeaters themselves are powered directly from the ring cables. This enables the attached devices to be switched off or removed, still leaving the repeater functioning. It is possible to use other media, including fibre optic cable, which enables greater distances between repeaters to be realized. However, in this case the line powered system is not able to be implemented directly as with twisted pair or coaxial cable.

The transmission rate for the original Cambridge ring is 10 Mbps, but the ring overheads and number of slots accommodated in the ring reduces this to approximately $4/(n+2)$ Mbps, where n is the number of slots circulating in the ring[6]. Thus the transmission rate can never be greater than 1.3 Mbps with $n = 1$. This reduction is usable transmission rate is due in part of the high proportion of bits contained in the frame for data management and control. A Cambridge ring packet is shown in Fig. 7.4. Two 8-bit data bytes

(a)

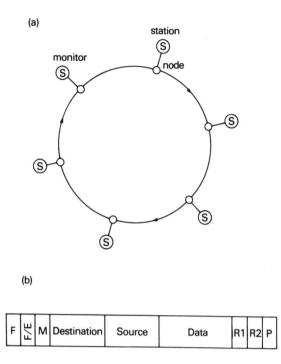

(b)

| F | F/E | M | Destination | Source | Data | R1 | R2 | P |

Fig. 7.4   (a) The Cambridge slotted ring.   (b) Packet format.

are accommodated in each ring packet. The remaining bits of the 38-bit frame are allocated as shown in Table 7.1.

The basic service provided by the Cambridge ring is an addressed packet and is not entirely suitable for carrying data directly. An alternative packet format has been applied by University College, London, to permit the ring to carry data at 2 bytes per packet for terminal support only. A virtual circuit is established between a terminal multiplexer and the computer interface in terms of two separate channels for duplex operation. The modifications to the Cambridge ring packet include channel numbers which identify the destinations for the packets carried on these channels[7].

In a more recent development referred to as the **Cambridge fast ring** (CFR), the clock rate is increased to 60 Mbps[8]. This enables more data to be contained in each slot and a greater number of packets circulating in the ring. In the CFR each slot contains 256 bits of data and over 65 000 different addressable destinations. This multi-address capability is valuable in allowing the CFR to interconnect numbers of small LANs without the necessity of using several bridging nodes. Several commercial implementations of the Cambridge ring are available, including Logica's Polynet, Xionic's Xinet and Tollec Dataring, all of which use longer slots than the original Cambridge ring.

Table 7.1

Cambridge ring packet content

1	Flag bit to show beginning of slot
2	Full/empty bit
3	Monitor bit
	Bits 2 and 3 together indicate:
	11 Sender has just transmitted a slot
	10 Set by monitor station to show that slot has passed. If the monitor reads this it knows that an error has occurred and deletes the slot
	01 Set by monitor in an empty slot
	00 Sender empties slot
4–11	Destination address
12–19	Source address
20–35	Data fields
36–37	Response field used as:
	11 Set by sender
	10 Set by receiver to indicate slot was rejected
	01 Set by receiver if slot accepted
	00 Set by receiver to indicate busy
38	Parity bit. Reset by each repeater as data passes

The error rate with the Cambridge ring system is extremely good (1 error bit in 10^9) so that none of the systems developed from it have found it necessary to make provision for error detection and control. The packet length is limited compared with Ethernet, so that more packets are needed to transmit a given length of data. Control is, however, much simpler and contention between stations cannot occur. Unlike Ethernet, the slotted ring is not limited to one message carried by the network at any given time. Since it is possible to have more than one slot circulating in the ring, interactive access becomes faster and more than one user can use the network at the same time.

7.4.3 The PLANET system

The ring systems offer possibilities of enhancement to improve reliability. One of these is to provide two rings, all critical resources being attached to both. This is, however, an expensive solution. A more acceptable approach is that marketed by Racal-Milgo in their PLANET (Private Local Area Network) system.[9] This is an empty slot ring having a number of features not

[9] PLANET: Racal-Milgo Ltd.

included in the original Cambridge ring. The essence of the scheme is shown in Fig. 7.5a. It consists of a double coaxial ring with one or more controlling directors (rather like the monitors in a Cambridge ring). In normal operation the director contains the two ends of a loop system as shown. In the event of a fault on one of the nodes the input/output connections to that particular node are connected together with the director, joining the loop ends as shown in Fig. 7.5b. The faulty node is thus simply removed from the network and with it any possibility of the fault effecting the system.

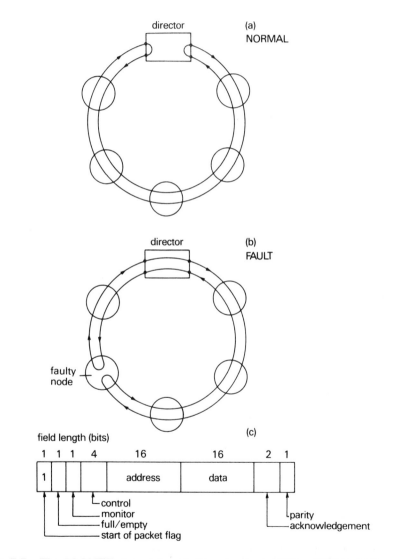

Fig. 7.5 The PLANET ring network (by courtesy of Racal Milgo Ltd).
(a) Normal operation; (b) fault operation; (c) packet format.

The ring packet format contains extra bits to indicate the way in which the address and data fields are used. This permits the PLANET system to implement a number of different higher-level services not available with the Cambridge ring. These are:

a **Broadcast calls**. A single message addressed to all stations on the ring.

b **Fixed calls**. Two ring users connected permanently through the director.

c **Designated calls**. Designated calls from one particular station to another arranged by one of the stations.

d **Switched calls**. Connections existing only for the duration of a call.

e **Diagnostic calls**. Network operation checks initiated by the director.

The slot size used is 42 bits of which only 16 bits are used for data (Fig. 7.5c). Data transmission is at 10 Mbps and the system can support up to 500 end devices.

7.4.4 Star ring architecture

Insertion of a new station on a ring system is not simply a matter of adding a tap connection as in a bus network. The ring needs to be broken and the new station, with its additional cable connections, made to complete the ring once more. Further, a fault on any node can render the complete network in-operative unless arrangements are made for the faulty node to be bypassed with connections made between its links to adjacent nodes. Both of these problems can be overcome at the cost (in many cases) of additional cabling through the use of a **star ring architecture**, shown in Fig. 7.6⁹. Here the connections from all the repeaters link through a single **ring wiring concentrator**, rather like a star network, although in this case the connections are actually from node to node through the concentrator and not radiating out from a central hub node. Because of this centralized access, it is a simple matter to isolate a fault and to insert new node connections. It also facilitates routing between a number of separate rings.

The principle of the star ring has been adopted in several ring systems, the most important of these being the **token ring** which employs a simplified form of data relaying known as **token passing**.

7.4.5 Token passing

The kind of bus operation we discussed earlier gives a high resilience to station malfunction and good physical access to the network. The empty slot ring used in the Cambridge ring is less resilient, but it is easier to implement

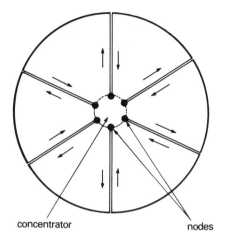

concentrator   nodes

Fig. 7.6   Star ring architecture.

and the protocol does allow guaranteed access. However, neither of these systems offers sufficient performance in terms of delay throughput for many modern applications where **token passing** would be used[10].

This alternative ring structure was first described by Penny and others in 1979[11]. Its operation is shown by reference to Fig. 7.7. In such a network a number of communicating nodes are arranged in a ring around which a single token circulates. At any time when no message is being sent the ring will contain only a series of bits corresponding to the token. The token length will be short compared with the slot frame in a slotted ring, since its only function is to indicate whether or not the ring is free to accept data. Hence the transmission delay introduced by the token can be quite small.

In operation, no node is permitted to transmit unless it is in possession of the token. To transmit a message a node must first remove the token from the ring and insert its message, appending a new token upon completion. In this manner all the nodes have the same chance to transmit. When there is a high demand for transmission the nodes will, in effect, form an orderly queue, each awaiting its turn. Since, in general, messages are long compared with the ring length, the efficiency of the ring is high. However, the token ring does nothing to improve resilience of the system, since a station must include a repeater to pass the token (and message) on round the ring and the failure of any one of the repeaters will lead to total network malfunction. The token ring then requires carefully designed fault management techniques. The two major problems that can be present in a given system are no token circulating or a persistent busy token.

In many systems one station is designated as an active token monitor and is able to record its monitoring activity by changing the state of a monitor bit contained in the token frame. To detect the lost-token condition a time-out is

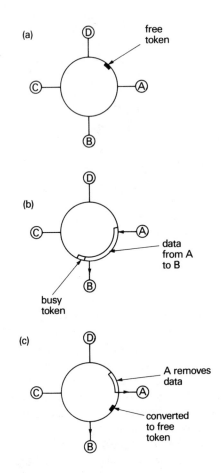

Fig. 7.7 Token-passing.

initiated greater than the time required for the longest frame to circulate the ring. If no token is detected in that time, the monitor deletes any residual data in the ring and issues a new token referred to as a 'free token'. To detect a persistent busy situation the monitor sets a monitor bit to 1 on any passing busy token. If it detects a busy token with the bit already set, is knows the transmitting station has failed to delete its frame and can take the necessary action by changing the busy token to a free token.

The principal advantage of a token ring is that traffic can be regulated by allowing stations to transmit differing amounts of data when they receive the token, or by setting priorities so that higher-priority stations have first claim on the circulating token. In addition, as mentioned previously, a token travelling around the ring and consisting of only a few bits will be propagated considerably quicker than the more lengthy circulating slot.

7.4.6 ICL MACROLAN

A modified form of the token-passing ring called MACROLAN, designed to improve the resilience of the system, is implemented by ICL in its latest series of main frame computers[12]. This applies a particular form of polled star ring architecture shown in Fig. 7.8 (included by permission of International Computers Ltd).

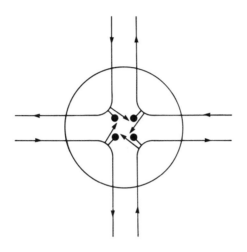

Fig. 7.8   Macrolan port-ring switch (by courtesy of R. W. Stevens and International Computers Ltd).

In this system, optical fibre transmission links connect up to 150 stations together via one or more port switches at a switching rate of about 48 Mbps. Each station is connected to its port switch via two optical fibres. One fibre is used for transmission and the other for reception. The port switch is arranged to monitor continually the activity of attached stations, switching out those stations which are inactive and thus considered inoperable.

A station initiates a message by sending a 'go ahead' (GA) character to the attached port switch. This responds by sending a similar character to the next operable link. This process continues until the character is received at the transmitting station, having completely traversed the ring. The receiving station then changes this to a 'start of frame' (SOF) character and transmits its message. A station not wishing to transmit simply returns any 'go ahead' characters unaltered back to the port switch.

On receipt of an SOF character the port switch changes to a different mode. Instead of circulating the GA character as before, it broadcasts the SOF and message simultaneously from each port. The port switch takes no action on completion of the message but simply awaits for a new GA character to be generated. This is initiated by a station wishing to transmit and a similar polling operation will then commence.

A major feature of the MACROLAN system is that message acknowledgement can be carried out at a low protocol level using a special field outside the message frame for this purpose. Stations receiving such a message pass an acknowledgement (ACK) character back to their attached port switch. The switch having received ACKs from each station then sends a further ACK character to the transmitter. Thus no action above link level (see next Chapter) is required to complete this operation.

7.4.7 IBM token ring

The IBM token ring is essentially designed to link IBM PCs together. It is a baseband system operating at 4 Mbps and uses twisted pair media. The data circulating the ring is encoded to the Manchester bi-phase format with the delimiters required in the frame sequence identified through deliberate violation of this code. (We discussed this technique earlier in connection with HDB3 coding in section 4.2.3.) Data frames contain a 4-byte destination and a 4-byte address. A control byte is added to indicate whether the data are synchronously or asynchronously encoded. Control for the network resides in a multistation access unit (MAU) which uses a PC running LAN software to act as a file server operating under the PC-DOS operating system and using a special IBM NETBIOS chip (see section 7.6). The connection to the network follows the starring architecture shown in Fig. 7.6, where the ring wiring concentrator is the MAU, permitting interconnection with up to eight PCs or to further MAUs.

7.4.8 Jitter in token-passing networks

A difficulty that can arise with ring networks, particularly token-passing systems where the message length can be small, is that the message needs to be repeated at each node station and is thus subject to any phase error or **jitter** arising during the regenerative process. This is a cumulative process since, as the message traverses the ring, it goes through many similar processes at each repeater. Unlike a typical digital transmission system which tends to exhibit random data traffic, ring systems carrying computer-generated traffic may have long strings of repetitive pattern data (e.g. a series of 1s). This kind of traffic will produce an amplitude of jitter which is a function of the message statistics as well as the characteristics of the transmission media.

During operation of a token ring network, the tokens and messages passing through the network are equalized, regenerated and re-timed at each node. Inevitably, because of non-ideal circuit implementation, each re-timing operation adds jitter to the data so that in a large ring this becomes the principal transmission impairment. The accumulation of jitter is proportional to the square root of N, where N is the number of nodes included in the transmission path. A phase-locked loop system may be applied to reduce this source of error[13].

7.5 REGISTER INSERTION

A third form of network interconnection which we need to consider operates in a manner similar to the slotted ring. This is known as **register insertion** and was mentioned earlier in connection with the MULTILINK and SIMPLENET systems (section 7.2.2.). It is a technique particularly suitable to ring-shaped LANs. The mode of operation is illustrated in Fig. 7.9a. When a device attached to a station has information to transmit, it loads this into a shift register. To transmit the information the register is then switched in series with the ring connections at a node so that it forms part of the data transmission path through the network (Fig. 7.9b).

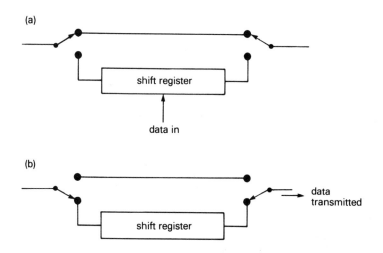

Fig. 7.9   Register insertion.   (a) Loading information;   (b) transmitting information.

The switching occurs whenever there is a convenient gap between other packets travelling round the network. The register stays in series with the ring so that all the packets of data may be diverted through the register. When the packet originally transmitted by the device at that location returns to it and is completely stored in the register, the register is then switched out of circuit. This is, in principle, fairly simple, but becomes much more complex in a practical system since the speed at which the register is switched in and out is high and the destination device must be able to read the data accurately and indicate that they have been completely received.

7.5.1 The SILK system

One of the earliest commercial register-insertion schemes was that developed by the Hasler Company of Switzerland and known as SILK (System for Integrated Local Communications). This is shown in Fig. 7.10. The shift register is connected in series with the ring as before, but is variable in length, determined by the position of the pointer shown in the diagram. Initially the register is empty and the pointer is at its minimum position at the left of the diagram. When the device has data to transmit it loads them into the transmit register one byte at a time. (The data in the ring take the form of a minipacket of up to 16 bytes.) At a suitable gap between the packets circulating the ring, the switch Y connects the transmit register and the data are transmitted. Incoming data from X now go into the shift register and the pointer progressively moves to the right (towards maximum) as the data fill the register. If the incoming characters are 'idle' characters, sent when no data are being transmitted, then the pointer remains at the minimum end of the shift register. Upon reception, the device addressed by the data packet recognizes its own address and causes switch X to operate, diverting the incoming data into the receiver register. When the end of the packet is reached, the shift register is then switched back into the ring. In order to maintain the flow of packets around the ring, whilst incoming data are thus diverted into the receiver register, either data are supplied from the shift register, if this has data to transmit, or the idle generator sends idle characters along the ring.

A feature of the SILK network is that it is capable of sending both digital data and digitized telephony information over the ring and was one of the first LANs capable of offering such an integrated service[14].

7.6 NETWORK SERVERS

The recent development of the PC with its semi-standard and modular construction has led to a number of networking systems becoming available in which the PC takes a central role in connecting to other PCs and to a network. Two factors have led to this. One is the possibility of interface and controlling hardware being contained on an interface card to be inserted into a free card slot within the PC, and the second is the ability of such systems to run under the PC-DOS or MS-DOS operating systems.

Strictly speaking, the majority of such systems are actually **file-server** systems and act as file managers for disc-based operating systems. Examples

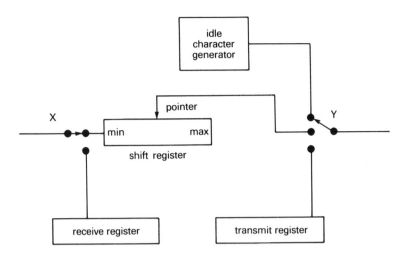

Fig. 7.10   Hasler SILK network.

are Novell's NETWARE and IBM's PCLAN programs. Such products give a local area network facility by providing security, access rights, utilities etc. for a variety of LAN hardware systems such as OMNINET, IBM token ring, ARCNET and many others.

Other systems provide local area networking through interconnection of a number of PCs which contain NETBIOS — an extension of the IBM PC's BIOS (basic input/output system) chip, the heart of the modular PC design. NETBIOS comprises a group of programs in ROM. These provide four alternative configurations for each connected PC:

a redirector,

b receiver,

c messenger, and

d file/print server.

The redirector configuration allows a PC to share resources and to send messages to other linked computers. The receiver configuration adds the capability to receive messages from other users and to display, print or store them in a file. The messenger configuration allows the PC user to switch between a local application and the network program and includes a screen editor. Finally the file/print configuration establishes the shared file/printer server for any PC with a hard disc. It gives each network user several access rights: read only, print only, read/write to a named disc or software directory.

Operation of the NETBIOS system uses the IEEE 802.2 LLC sublayer of the OSI model which is discussed in the next chapter. Control is through a number of specific commands which are additional to the set of commands found in PC-DOS or MS-DOS. One such system is LAN-25[10] which operates as a gateway, permitting linkage to an external public or private packet-switched network conforming to the X25 protocol. In use the interface card software system extends the normal PC-DOS control to include the IBM NETBIOS commands and has the capability to establish up to 32 virtual calls across the X25 network.

A recent development is a more complex chip set, the Texas Instruments TMS380 unit which provides a set of IEEE 802.5 compatible protocol services and allows adaption or integration with a number of standard OSI-layered LAN and WAN systems. For example, a TMS380 network integrator can implement the physical and medium access control layers to match the function of the IBM token ring or the Xerox network system (originally developed for an Ethernet LAN). Thus, in a suitably designed plug-in inter-face card incorporating the TMS380, a direct connection to a token ring or Ethernet compatible system becomes possible.

As with the NETBIOS chip, the TMS380 provides a number of separate programs, but these are more complex than with NETBIOS and are contained as a series of interconnected ROM chips shown in Fig. 7.11 (included with permission from **Mini-Micro Systems Journal** and Texas Instruments Inc.). These are seen to comprise a five-chip set — a system interface, a communication processor, a protocol handler, a ring controller and a ring interface. In operation, the TMS380 acts as an interface between the lower layers of the protocol to be used with the PC (i.e. token ring, Ethernet, or other systems) and the PC's own operating system. The interface can be set up and programmed through software read into the PC. This means that the system can act as a file server and gateway to a variety of different network systems by a change of stored software. It is particularly suitable for providing compatible connection with a token ring LAN and other systems[15].

7.7 BROADBAND SYSTEMS

All the LAN systems discussed above are **baseband** systems, that is, the entire transmission bandwidth is employed for the signal being conveyed through the network. If multiple message are to be transmitted, they are arranged to follow one another sequentially using a form of **time division multiplexing**. An alternative way of using the transmission bandwidth is to apply the principles of **frequency division multiplexing**, described in Chapter 2, by transmitting simultaneously a number of messages, each at a low trans-mission rate, determined by the way in which the available transmission

---

[10] LAN-25: The Software Forge Ltd, UK.

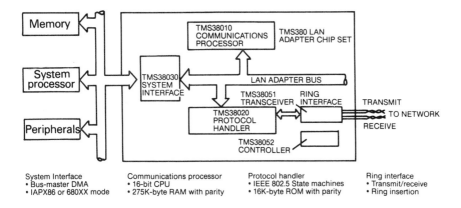

Fig. 7.11 Texas TMS380 OSI protocol chip (by courtesy of Texas Instruments Inc.).

bandwidth is subdivided amongst the several channels used. In network terminology this is referred to as a **broadband** system. Here an analogue carrier is used to support an FMD system and individual digital signals are applied to the carrier through the use of modems. A pair of modems, tuned to the same carrier frequency are used for each channel. Alternatively a switched modem, known as a **frequency-agile modem** may be used to relate the modem frequency to a particular transmission. This adjustment of the received modem frequency to the transmitter modem frequency is carried out automatically. Where multiple access on a given channel is used, then some form of access control is needed as with baseband systems.

Broadband systems are capable of transmitting greater distances than baseband systems without the use of a repeating station. Some tens of kilometres are possible, since the analogue signals that carry digital data can propagate greater distances before noise and attenuation affect the data. Fig. 7.12 shows a typical broadband system. It is inherently a **uni-directional** system since analogue amplifiers need to be used on the media and can propagate only in one direction. As a consequence, two data paths are needed to support duplex operation. These may be two physically separate transmission paths or a single path conveying two signals transmitting on separate frequencies. The paths are joined at a point on the network, known as the **headend**. All stations transmit on one path towards the headend and receive on a second path away from the headend or, in the separate frequency system, the headend arranges for a frequency translation to take place between the transmit and receive frequencies.

Although the configuration for a broadband system is inherently more complex than a baseband system, it benefits by being able to make use of some of the components which have been developed for community television transmission (which operates as an analogue service over a similar range of

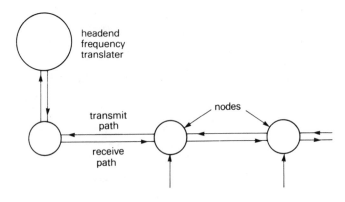

Fig. 7.12 A broadband system.

frequencies). These include the coaxial cable, terminators, amplifiers, controllers and directional couplers. This last provides a means for dividing one input into two outputs or for combining two inputs into one output.

7.7.1 Broadband LAN systems

A number of broadband systems are in use for local area networks. These are much more suitable for some applications than baseband systems. In particular, broadband transmissions are much less susceptible to interference from electrical machinery and are hence popular for industrial process control such as the MAP systems discussed in Chapter 9.

The bandwidth provided is greater than that found in baseband systems and consequently the total throughput is greater than in practically every other viable system. The ability of broadband systems to handle both analogue and digital information equally easily is also useful. In the following several currently available broadband systems will be considered.

7.7.2 MITRENET

MITRENET[11] uses ordinary TV coaxial cable and transmits in a frequency band of 300–400 MHz. Since available television equipment operating at this frequency operates at a channel bandwidth of 6 MHz, much wider than is needed for data transmission, time division multiplexing was orginally applied to permit multiple channel transmission along the network coaxial cable. This broadband system applies separate 'go' and 'return' paths through

---

[11] MITRENET: The Mitre Corp., USA.

the use of separate transmit and receive frequencies which are interpreted at the headend (see Fig. 7.12).

A later development in the system is to allocate channels on a frequency division basis but still using the TV cable technology. The network protocol is a form of CSMA/CD technique applied in a frequency-modulated context. Very high usage devices requiring continuous channel capacity for long periods can be allocated separate frequency channels which are not shared and, in fact, look like private leased lines on a WAN PSDN system. Analogue transmission (i.e. speech and television) can use the same cable, since these services can also be given dedicated bands within the total frequency bandwidth.

7.7.3 WANGNET

The WANGNET system[12] applies the MITRENET ideas but uses a two-cable system instead of using separate frequency channels for the 'go' and 'return' paths. This enables the full frequency bandwidth of 300–400 MHz to be used for each direction of transmission. This total bandwidth is divided by WANGNET into three separate bands used for different purposes:

a Wang band: for communication between Wang's own processors and equipment operating in a shared mode and using a form of CSMA/CD protocol. The total transmission rate is 12 Mbps.

b Interconnect band: to interconnect devices from other suppliers and terminals using either dedicated 64 kbps point-to-point channels using fixed-frequency modems, or multi-channel data switching applying frequency-agile modems to establish the channel in use.

c Utility band: which is split into separate channels for information not generated by the computer; for example, analogue or digital speech and television.

7.7.4 ARCNET

ARCNET[13] is a comprehensive network designed to permit interconnection for a number of network environments in addition to its principal function as a broadband network. In this capacity it uses dial-up modems, operating at 9.6 kbps and thus makes use of the PSTN as the network carrier. Control is exercised by a small node computer in which the operating software is entered through a microdisc reader. This enables considerable flexibility to be

---

[12] WANGNET: Wang Laboratories Inc., USA.
[13] ARCNET: Datapoint Inc. USA.

obtained through loading alternative software held in disc storage for different operating nodes. For example, the network channels can be connected directly to a packet-switching line using X25 protocol. To achieve this a packet-assembler/disassembler forms part of the controlling equipment as in the Arpanet TIP. Connection to an ARCNET token-passing bus network operating at 2.5 Mbps can also be made from the node. (ARCNET was one of the first token-passing networks to become available commercially.) In the token-passing ring the nodes are arranged to form a logical ring to a controlling hub node and several such rings can be connected together via these hub nodes.

The broadband network is connected through coaxial cable with a total data rate of 2.5 Mbps. This permits up to 255 nodes to be connected, with line lengths between nodes of up to 6 km. To extend the connection to individual stations (printers, terminals, PCs etc.) the broadband controller may be connected to a number of sub-controllers acting as demultiplexers and data formatting devices.

7.7.5 Spread spectrum network

An interesting alternative to baseband time division multiplexing used with most bus networks, is spread spectrum code-selective multiplexing devised by C. T. Spracklin and others[16]. This makes use of signal processing concepts such as orthogonality and correlation to maintain the identity of separate channels within the multiplexed data stream[17]. In operation, a channel is associated with a given rectangular pulse code sequence used as a carrier signal for the data in place of the more usual sinusoidal carrier signal. The code sequences (which are different for each channel) must have an orthogonal relationship to each other since the method of decoding used demands a minimal cross-correlation coefficient between them.

The data for a given channel is initially added modulo-2 with the code sequence chosen for that channel. This constitutes the modulated data for transmission. Data for other channels are similarly added to their respective code sequence to produce further modulated channels. The summation of all these modulated digital channels is then transmitted along the communication media.

Since the set of codes is an orthogonal one it is permissible to interleave a number of coded channels in this way without losing their individual identity. At the receiving end, to which a number of channel nodes may be connected, each node station is aware of the code sequence related to its own channel and is thus able to cross-correlate the incoming interleaved data with this code sequence in order to extract the modulated message data. A further stage of modulo-2 addition is required to derive the actual message from the recovered modulated signal.

Contention between channels for the network is reduced if the transmitter, rather than the receiver, is allocated the given code sequence. This implies that the receiver must decode to a series of code sequences instead of to just one and hence requires a number of cross-correlating units operating in parallel. Further details of this interesting technique may be obtained from the references given above.

SUMMARY
Chapter 7

The limited geographical area for a LAN permits some simplification in protocol design since routing is no longer required and compensation for the media characteristics is of less importance. One consequence of this is that the transmission rates can be considerably higher than those found in a WAN.

The choice of LAN technology is now extensive, permitting a mode of operation to be selected to suit the media and topology chosen. This has led to two broad divisions of networks: **linear systems**, typified by the bus or tree topology, and **ring systems**. The former presents a problem of possible simultaneous access to the network by more than one station leading to a collision, corrupting the information being transmitted. The ring systems, on the other hand, require a mechanism to be devised to permit a station to access and recover data from the ring and to prevent continuous transmission around the ring once the message has been successfully sent and received. A number of different techniques and architectures have been developed to solve these problems. These make use of alternative protocols discussed in the next chapter.

A further dichotomy in network design is to refer to a network design as a **baseband system** in which the entire bandwidth of the network is applied to the transmission of a single data stream, with a number of channels accommodated through time division multiplexing, or to **broadband** transmission where the available bandwidth is divided between separate data streams permitting a number of channels to be accommodated in the network through frequency division multiplexing. Several such broadband network systems are available, some of which permit services, other than data transmission, such as speech and vision information to be carried through allocated bandwidth channels.

PROBLEMS
Chapter 7

P7.1 Describe the effects of a complete failure of a node in the operation of the following network configurations:

    a a bus,

    b a ring,

    c a star.

P7.2 A slotted ring and a token ring, each having 200 stations on the ring, exhibit the following characteristics:

    length of slot = 50 bits,

    length of token = 8 bits,

    data rate of the slotted ring = 10 Mps,

    data rate of the token ring = 5 Mps,

    Both rings have the same overall length of 5 km.

    Calculate the maximum number of slots or tokens which can circulate in each network.

    (The propagation velocity can be taken as 2×10^8 m/s.)

P7.3 An Ethernet having a transmission rate of 10 Mbps has a total network length of 1.5 km and a propagation speed of 200 m/μs. The data packets are 512 bits long and include 32 bits of header, checksum and other overheads. An acknowledgement packet of 32 bits is sent after a data packet is received. Find the effective data rate for the information contained in the packet (i.e. excluding overheads).

P7.4 A Cambridge ring has a transmission rate of 10 Mbps and an overall length of 10 km. The slot length is 40 bits. The number of repeaters in the ring is 350, each of which introduces a 1-bit delay into the transmission. How many slots can be included in the ring? (Consider the speed of signals along the line to be 2×10^8m/s.).

P7.5 **a** What is a baseband LAN?

What is a broadband LAN?

b How is duplex operation obtained in a broadband system? Why is bi-directional transmission not possible in the same way as with baseband transmission?

P7.6 What are the advantages of using a star ring architecture in a computer network? What are its disadvantages?

8

PROTOCOLS

To enable digital devices and computers to communicate with each other there must be rigorously defined **protocols**, i.e. sets of rules about how control messages and data messages are exchanged and how the former can control the communication process. It is also necessary to define equally carefully the formats of these control messages to remove any possibility of ambiguity[1]. The key elements of a protocol may be given as:

a Syntax: includes formatting of the data, coding and defining signal levels.

b Semantics: includes synchronization, control and error handling.

c Timing: includes data rate selection and correct data sequencing

We will meet all of these elements in our examination of network protocols.

The formats and protocols used for data exchange can become quite complex, particularly where the intention is to link together several different kinds of devices and different manufacturers' computers. Manufacturers often use different data formats and data exchange conventions even within their own product lines. A software package required to provide a general solution to the problems inherent in trying to pass information between any of these possible systems would be far too complex to attempt.

This complexity can be reduced by the adaption of a hierarchy of protocol standards, provided they are accepted universally. It must be said that, at present, the process is by no means complete although certain areas of data interchange are being agreed internationally. These standards are based on the general principles of **protocol layering**, and we begin our discussion with an explanation of how the acceptance of these techniques can simplify the problem.

8.1 PROTOCOL LAYERING

The principle of using a number of different levels (**layers**) of software is well known from the general historical development of computing software, as is shown in Fig. 8.1. Initially, the user accessed the computer directly, using machine code consisting of the basic 0 to 1 logic levels required of the computer (layer 1). Later, assemblers and compilers were designed, permitting access through programs consisting of alphanumeric mnemonic codes and higher-level computer languages (layer 2). Operation of the computer, particularly input/output requirements, was made simpler by the use of an operating system (layer 3) and finally the user is seen to organize his information much more easily with the aid of a data base management system (layer 4). Thus the software to operate the computer is seen to consist of a number of 'layers' of coding, each dependent on the layer beneath it. The user's application program can then be written to address the actual problem to be solved, leaving the business of data sorting, calculation and presentation to the basic machine hardware under the control of layers of software coded instructions.

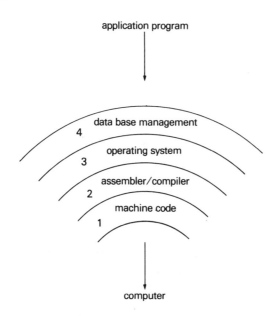

Fig. 8.1 Principal of software layering.

The same principle may be used for the process of communicating between data sender and receiver. The communications functions are similarly partitioned into a vertical set of layers. Each layer performs a related subset of the functions required to communicate with another system (computer or end device) with each layer relying on the next lower layer to perform the more primitive functions that are needed and to conceal the actual operating details of those functions. In turn, the layer provides services to the next higher layer. Ideally, the layers should be defined so that changes in one layer do not require changes in the other layers. In this way, the communication problem may be divided into a number of more manageable subproblems which can be defined in such a way that the total combination can have wide applicability.

One of the earliest networks making use of protocol layering for subdividing the communications operations was Arpanet. This uses two levels, a higher level for user function processes and a lower level, split into three sublayers for communication. These lower sublayers correspond to the IMP–IMP, IMP–host and host–host protocols (see section 6.5.1). The higher layers include Telnet, a general character-oriented communications protocol, an interconnection protocol (ICP) which provides for remote user process initiation, a file transfer protocol (FTP) to transfer data between two hosts, a remote job entry protocol (RJE) to allow a user at a host to run a batch job, and lastly an electronic mailbox protocol which permits plain language messages to be conveyed between end users.

8.1.1 The ISO seven-layer standard

The most important and widely accepted of these layer systems is the **seven-layer protocol system** proposed by the International Standards Organization (ISO). This system is one aspect of an interconnection philosophy known as **open systems interconnection** (OSI)[2]. **Open systems** are simply systems that are capable of interconnection by virtue of each system having implemented a common set of protocols.

OSI refers to communications between computer systems which are enabled to exchange information freely irrespective of type and manufacturer by virtue of their mutual adherence to a set of standards. The standards may be described by references to an architectural model of a communications system, the **ISO seven-layer reference model** which has now reached international agreement through the work of ISO (whose initials are given to the system), the International Telegraph and Telephone Consultative Committee (CCITT), the American National Standards Institute (ANSI), the British Standards Institute (BSI) and several other bodies.

This technique reduces the complexity of the software systems being modelled so that each layer is responsible for one particular aspect of the communication problem. This is shown in Fig. 8.2. The seven layers of protocol are:-

1 **Physical layer**: This layer describes the nature of the physical circuits which allow the transmission of a stream of data bits between the two ends. It is concerned with fundamentals such as pin connections and

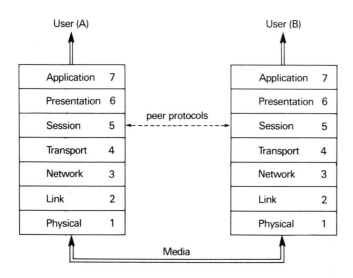

Fig. 8.2   ISO seven-layer communication model.

electrical voltage levels. The most common Layer 1 standard in use today is RS-232c described in Chapter 2 as a method used to connect a DTE to a DCE, such as a modem.

2 Data link layer: This layer defines protocols for transferring messages between the DCE and DTE, ie host-to-network and vice versa. It also performs error detection and correction for errors occurring in the data transferred and carries out data synchronization. In effect, it converts a simple physical, but possibly unreliable connection, into a tested and error-free digital connection between two locations.

3 Network layer: This layer supports network connections between two hosts communicating over a network and also allows multiplexing of several logical communications channels down the same connection. It converts the reliable digital connection provided by layer 2 into a multi-node network exchange of data. The network layer thus permits the setting up of an actual data transfer across the network.

These first three layers are the **communication-oriented layers** and are applied, for example, in the CCITT X25 protocol which is used in a public packet-switched service.

4 Transport layer: This layer provides for the transparent transfer of data, e.g. files, between end systems which might organize their data somewhat differently. It relieves the transport users from any concern about *how* in detail data transfer may be affected and optimizes the available communications resources.

This layer can be viewed as a bridge between the **communication-oriented** lower three layers and the **application-oriented** upper three layers described below.

5 Session layer: This layer supports the establishment, control and termination of dialogues between application processes. It facilitates full duplex operation and maintains continuity of session connections. It also supports synchronization between users' equipment and generally manages the data exchanges.

6 Presentation layer: This layer resolves differences in representation of information used by the application task so that each task can communicate without knowing the representation of information used by a different task (e.g. different data codes). In effect, this means the ability to run a job on an 'alien' machine. Taken in conjunction with the top layer, the application layer, this can make the network completely machine-independent.

7 Application layer: This is the highest layer in the reference model and is the ultimate source and sink for data exchange. It provides the actual user information processing function and programs for application processes in the real world (e.g. airline booking, banking, electronic mail etc.).

Some useful **OSI terminology** is shown in Fig. 8.3. We refer to a given layer as the **(N) layer**, and the names of the activities connected with that layer are also preceded by (N). Within a given layer there are one or more active **entities**. Entities are abstractions of whatever is required to carry out the function of a given layer. They may correspond to one or more software or hardware modules or even a complete microprocessor implementation. In essence, they are processes capable of sending or receiving information. An **(N) entity** implements functions of the (N) layer and also the protocol for communicating with (N) entities in other systems. An example of an entity would be a single process in a multiprocessing system. Or it could simply be a subroutine used to carry out a repeated function. There might be multiple identical (N) entities within a given layer, if this is convenient or efficient for the system.

Each entity communicates with entities in the layers above and below it across an **interface**. The interface is realized as one or more **service access points** (SAPs), which function in a similar manner to hardware ports. (This process can be considered as a form of multiplexing between layers as we shall see later.) The (N-1) entity provides **services** to an (N) entity through the use

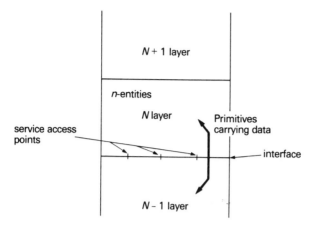

Fig. 8.3   OSI terminology.

of **primitives**. These provide the interactions between two layers and are the sole means by which adjacent layers in the same system communicate and exchange parameters. A primitive specifies the particular function to be performed and is used to pass data and control information across the interface between layers.

This process allows for each layer the definition of a protocol which controls how the **peer entities** (i.e. entities in corresponding functional layers, shown in Fig. 8.2) communicate with each other. The operation of a simple communication process based on OSI principles is illustrated in Fig. 8.4 for the case where an application X has a message to send to an application Y across an intervening communications medium. The message is transferred to an application entity in the application layer. Here a set of control bits, referred to as the **header**, is appended to the data to provide the required information for the corresponding peer layer 7 protocol in application Y. The original data, plus the header, are now passed as a unit to layer 6. The presentation entity treats the whole unit as data, and appends its own header for action by the corresponding peer layer 6 and passes this to the next lower layer. This process continues down through to layer 2, which generally adds both a header and a trailer. This layer 2 unit, called a **frame**, is then passed by the physical layer 1 to the transmission medium. When the frame is received by the target system, the reverse process occurs. As the data ascend through the different layers of protocol, each layer strips off the outermost header, acts on the protocol information contained therein, and passes the remainder up to the next layer until, in the final application Y, the original message is acted upon.

Development of protocol standards to meet all seven layers of the model is not yet complete in terms of international agreement with layers 5, 6 and 7 the least defined.

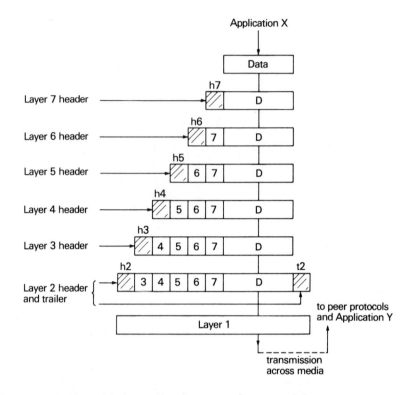

Fig. 8.4   Communication using the seven-layer model.

8.2   THE COMMUNICATION LAYERS

The formats of the communication layers have now reached fairly general agreement and find their most widespread acceptance in the **HDLC** and **X25 standards** which we now need to consider.

8.2.1   High-level data link control

Many of the LAN protocols in use today are based on the **high-level data link control** (HDLC) protocol developed by ISO in the early 1970s[3], but are generally modified to improve the somewhat complex procedures for error handling and flow control.

HDLC is a synchronous protocol, introduced briefly in Chapter 4, when we discussed synchronous and asynchronous communication. For reasons which will become apparent later, this and other similar protocols, such as Synchronous Data Link Control (SDLC) developed by IBM, are referred to as **bit-oriented protocols**.

HDLC has been specifically designed for communication in wide area networks so that a number of its functions, designed to assist routing or flow control, are not actually needed in a local area network. Note also that there is no necessity for the length of each character to be specified, since the bit representation of the data, in the form of characters, binary numbers, or decimal numbers, is contained wholly within the data field of a single frame.

HDLC provides three classes of procedure for network connection between adjacent nodes in a point-to-point communications system. The first of these is **asynchronous balanced mode** (ABM), applying full duplex communication where the two ends of the link are 'equal partners' in the data exchange (hence the 'balanced mode' connotation). Both can initiate and terminate a connection and send data, without prior interrogation, on an established connection. Secondly a **normal response made** (NRM) procedure can be initiated between a control device (computer) and a number of secondary stations. A third type of operation is **asynchronous response mode** (ARM), which applies half-duplex working on point-to-point lines. Here a primary station sends out commands and data. A secondary station then returns responses.

Although NRM and ARM have been used extensively in the past, the preferred mode of operation for computer-computer, point-to-point (and hence LAN) links is ABM, and it is this mode of operation which is discussed in this chapter.

In HDLC, all exchange of information, control, and acknowledgements is provided by formatting the data into rigidly defined frames. The **frame structure** required to support various link control procedures will now be considered, together with the commands and responses used for **data transfer** and the set of commands and responses required for different **modes of operation**.

The frame structure for HDLC is shown in Fig. 8.5 and consists of a number of separate fields:

a Flag: Used for synchronization and to indicate the start and end of a frame. The flag pattern 01111110 in the data field is avoided by **bit stuffing** as described earlier. (When there are no frames to be sent the flag sequence alone may be repeated indefinitely.)

b Address: Identifies the sending or receiving station.

c Control: Identifies the different modes of operation for the protocol (see below).

d Data: Contains the data to be transmitted.

e CRC: A frame check sequence field applying the 16-bit **cyclic redundancy check** described in section 4.4.2. The entire frame (i.e. the address, control and data fields) is included in the checksum calculation.

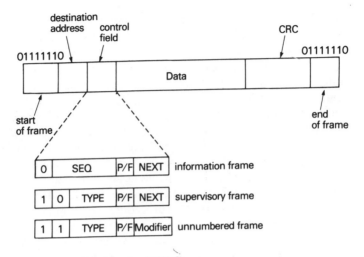

Fig. 8.5 Frame structure for HDLC.

Because HDLC has been defined as a general-purpose data link control protocol, a specific mode of operation needs to be selected when a data link is first set up. These are the ABM, the NRM, and the ARM described earlier and which are identified in the control field.

In order to provide the basic communication functions of the HDLC protocol, three different types of control field are defined:

1 Information frames: to carry the data.

2 Supervisory frames: to provide basic link control functions.

3 Unnumbered frames: to provide supplemental link control functions.

The contents of these three alternative control fields are shown in Fig. 8.5.

In the information frame the P/F (poll/final) bit is used by a primary station to obtain a response from its linked secondary stations. When a computer wants to transmit data to a terminal, it sends a poll message with the P/F bit set to 1. The terminal responds with a frame also containing a 1 in this position. More than one frame may be sent with only the last frame containing a P/F bit set to 1.

SEQ and NEXT fields are included in this frame to provide a technique for flow control and error control using the sliding window technique described in section 5.7.2.

The **supervisory frame** is the main mechanism for controlling the data exchange. Since the receiving end may be in varying stages of readiness to accept the data and the data previously transmitted may be subject to error and require retransmission, it is necessary to indicate these conditions within the frame. To do this four different types of supervisory frames are available. These are:

a Receive ready: used to acknowledge correct receipt of frames by noting the NEXT field in the information frame.

b Receive Not Ready: used to indicate a busy condition.

c Reject: used to indicate error in a frame or frames received and to request a retransmission of the faulty frame(s) numbered through the SEQ field.

d Selective reject: used to request retransmission of a single frame.

The class of control field referred to as the **unnumbered frames** are not given a sequence number and are used for several special purposes which are not considered here. Further information can be found in ref.[1].

8.2.2 X25 communication protocol

A second important protocol making use of the communication layers is the CCITT proposal X25 which was mentioned earlier in connection with System X, PSS, and other packet-switching systems for the PSDN. Whilst X25 finds its main application in a WAN it can also be used for some LANs, particularly where these need to link to an X25 WAN. First published in 1976 by CCITT, the X25 standard has undergone considerable refinement since then. In this chapter the details refer to the 1984 version, the 'Red Book'[4].

The specification for X25 provides an interface between the users' equipment and the communications network. These are the data terminating equipment (DTE) and data circuit terminating equipment (DCE) discussed in section 4.5. X25 implements a link access procedure similar to HDLC asynchronous balanced mode. In terms of packet transmission three basic types of service are offered:

a Virtual circuit operation: in which the DTE attached to a station can request a circuit connection which is then relayed to the station called. This is free to accept or reject the call. If accepted, connections are set up across the intervening nodes using HDLC protocol and the virtual circuit so formed is given a logical channel number. The data packets are then transported. To identify the station called, the network address identification follows another CCITT recommendation, X121 as shown in Fig. 8.6. The country code is set by CCITT, other codes are set by the network.

b Permanent virtual circuit operation: This is similar to virtual circuit operation but here a call is considered to be permanently set up and all packets directed to the same pre-arranged destination.

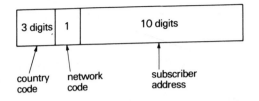

Fig. 8.6   X121 address format.

c Datagram operation: where neither call set up nor call clearing are required. One or more virtual circuits are reserved for datagram operation. Since these are unacknowledged transmissions, their integrity and order of packet delivery are not guaranteed. (Strictly speaking, datagrams are no longer supported under the 1984 CCITT recommendations (the red book), but many networks are functioning to earlier recommendations which do support datagram operation.)

The three communication procedures (layers) for X25 concerned with managing the data transfer are shown in Fig. 8.7. The X25 protocol is represented by the contents of the large box, and shows n processes engaged in n simultaneous virtual calls via a single X25 link.

There layers are:

a Physical level: The characteristics specified in X25 are also those described for other recommendations (e.g. X21 for linking synchronous devices — see section 4.5.4). These are widely used in data transmission systems and define the interface between the DTE, i.e. the customer's transmission equipement and the DCE, i.e. the equipment used at the connection to the network.

b Link level: This looks after the setting up, disconnection and overall management of the link. The structure is a balanced link access procedure, referred to as LAP-B, and is compatible with HDLC.

c Packet level. This specifies the packet types and formats, together with a procedure for establishing, clearing, and managing virtual calls and permanent virtual circuits. A data packet consists of a 3-byte header and a data field of length defined by the network (up to 4096 bytes).

In addition to the transmitted data packets various short control packets are used for setting up and maintaining calls across the network. Two of these are:-

a Call request packets: which can include up to 16 bytes of data.

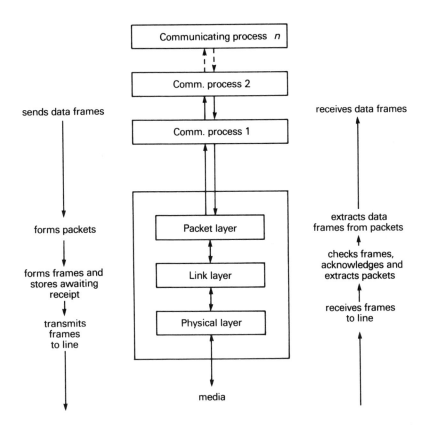

Fig. 8.7   X25 communication layers.

b Flow control packets: which do not contain user data and are only 3 to 4 bytes long.

X25 protocol can be used in a local area network; examples are Netskil[1] operating an Ethernet service and LAN-X25[2], a gateway service. It is not entirely suitable for this purpose, however, since broadcast facilities are absent and the X121 addressing convention does not provide the facilities for a large number of addresses which would be required in a LAN environment.

X25 is, however, an excellent vehicle for WANs and contains a good error control required for operation over noisy carrier frequency lines.

The specification of X25 protocol has now stabilized sufficiently to encourage the manufacture of hardware chips to control the working of the protocol

[1] Netskil: International Computers Ltd, UK.
[2] LAN-X25: The Software Forge, UK.

at its various levels. These are available from Fairchild, Intel, Motorola, Western Digital and others.

8.2.3 Packet assembler/disassembler

Since many DTEs in service are not able to support X25 directly, a protocol converter, known as a **packet assembler/disassembler** (PAD) is available to provide for simple synchronous non-intelligent character mode terminals. It is sometimes referred to as the **triple-X** PAD since its functional characteristics and interfacing protocols are specified jointly by three relevant CCITT recommendations, X3, X28 and X29. The main task of a PAD is to break up the digital message into packets for transmission through a packet-switched network and to reassemble the packets at the receiving end to form the original unbroken message. In practice, a device requiring to transmit information may present this in the form of a series of 8-bit characters (e.g. a data input terminal). It is very likely that the device will have no internal buffer capable of holding more than one character, so a buffer register is included within the PAD where the characters may be assembled until a complete packet is ready for transmission. Before transmitting the packet, the PAD will add the requisite headers and a trailer to conform with the requirements of X25 protocol, and only then transmit the packets to a device programmed to accept X25 data.

When the PAD receives a packet via the X25 network, it carries out a reverse process, removing the headers and trailer, and sending the characters one at a time to the start–stop terminal at its connected DTE.

Currently available PADs usually incorporate a concentration function which permits a number of DTEs to share a common X25 transmission path. The PADs applied to the UK JANET network, for example, are of this type and will accept 16 incoming lines for connection to the X25 network.

8.3 JANET PROTOCOLS

As discussed in Chapter 6, a set of layered WAN protocols was developed for the UK academic community at a time when ISO had only just begun work on the seven-layer reference model. These are used in the joint academic packet-switched network, JANET, linked all the University and Research Council sites and some Polytechnics. The protocols are related (in some cases quite closely) to the OSI recommendations and were listed in Table 6.4. Brief details of some of the most important of these standards will be considered in this section. An excellent review of their operation is given in ref.[5].

8.3.1   Network-independent transport service (NITS)

This is a host-to-host protocol intended to provide a common interface between the host application software and the lower-level network protocol. The protocol was seen as a comparatively simple method of transferring files and jobs between university campuses and the Research Council sites. It provides a set of primitives which are intended to be independent of the characteristics of the underlying network. Since the network makes use of the British Telecom PSS network, it follows that the NITS primitives conform closely with operation, at the X25 level so that it is not quite network independent, although it will operate between hosts on a number of other networks.

The primitives form the interface between NITS and the protocol at the next highest layer. They are used by this higher layer to carry out its command functions which are relayed by NITS to its peer protocol at the other end of the link where appropriate action is taken. In general terms, the primitives are used to establish and clear a call between hosts, to transfer data and addresses across the link and to look after other details such as resynchronization and priority calls. A full description of their operation will be found in ref.[6].

8.3.2   Network-independent file transfer protocol (NIFTP)

This, as the name implies, is a protocol for transferring files or any other document between host sites independent of network characteristics. NIFTP is implemented over NITS using the NITS primitives. Files to be transferred (which may originate from mainframe hosts, microcomputers and even from card readers and printers) are processed in three distinct phases:

An **initiating phase,** in which the host initiating the transfer first transmits a set of desired attributes to its peer process at the receiving host. If these are found compatible with the possible receiving host operations, the **data transfer phase** commences. The file is then transferred as a sequence of records of agreed length (one of the attributes), interspersed with a set of interim acknowledgements during the course of the transfer. A **termination phase** occurs when all the files have been successfully transferred.

The underlying NITS reports communication errors and failures to NIFTP which, itself, includes a set of failure and recovery procedures[7].

8.3.3   Job transfer and manipulation protocol (JTMP)

This is an applications protocol using NIFTP as the mechanism for file transfer and contains extensive job handling procedures. It is concerned with the processing of remote batch jobs to be run on other computers on the network. The user is able to define the machine to be used to process the job, the sort of actions required, e.g. compile, execute, list etc., what output is to be returned and where it is to be sent. Various accounting and control information is also provided with the output document.

Unlike the lower layers, JTMP functions by transferring processing tasks which are treated as files, using the carrier protocol NIFTP. Only JTMP is concerned with the contents of these files; NIFTP merely acts as the carrier. A number of format parameters, called **descriptors**, are added to define the operations required by the user. These are acted upon by protocol commands included within JTMP. The process does not go as far as incorporating the job description language required for the user's data. It is still necessary to include details of how the host computer is, for example, to process the actual data, e.g. compile and run a Pascal language program. It is, however, a true and powerful application protocol and completely machine independent[8].

8.3.4 Electronic mail transfer protocol (JNT)

This also is an applications protocol. The Joint Network Team (JNT) mail protocol is based on the distributed mail system used by Arpanet. Each user has a mailbox located in his local host which is identified within a general **mail server** address. This is applied by the user participating in the system to identify both the site addressed and the individual's mailbox location. Various pieces of information are required to establish the message status, e.g. date, subject, number of copies, reply etc., and these are included in a header contained in a presentation format. Compatibility with Arpanet enables messages to be exchanged between the two networks. Further details can be found in the 'grey book' describing the protocol[9].

8.4 LOCAL AREA NETWORK PROTOCOLS

In discussing LAN protocols we need to consider only the first three OSI layers, the communications layers. LAN protocols do not make use of the full set of ISO protocol layer recommendations, since end-to-end services of the kind provided in wide area networks to communicate between user processes are not required. It is sufficient to exchange a checked and reliable data packet stream between two devices on the network. Quite often a LAN is used simply to convey data to a shared printer or to transport formatted files in a word processor context where application services are not needed.

For LANs, the physical and data link layers of the ISO definition may be adequate, although often some of the attributes of the third layer, the network layer, are also required. As a consequence, layering protocols for LANs have developed into the three-layer system not directly equivalent to the three communication layers described earlier. The LAN layers consist of:

a **A physical layer** identical to the ISO layer 1.

b **A medium access control** (MAC) layer to manage communications over the link.

c **A logical link control** (LLC) layer to provide one or more service access points (SAP) which permit setting up of a form of multiplexing to handle multiple-source multiple-destination data.

In addition, the LLC layer assembles the data into a frame for transmission with address and CRC error checking bits, and carries out disassembly of the frame with address recognition and error checking upon reception. These last two layers taken together, are approximately equivalent to the ISO layer 2.

We need now to consider the LLC and MAC layers in a little more detail in order to see how the LAN protocols have been developed to form their own particular network standards.

8.4.1. The logical link control layer

The LLC layer is concerned with transmitting packet frames between two stations on a network having no intermediate switching nodes. In order to do this effectively it needs also to include logic to handle **multiplexing** and this is obtained by the concept of the **service access point** (SAP).

An example will show how this works. Figure 8.8 shows three stations attached to a LAN. Each station has its own address and each has a number of SAPs where data from a device can be entered. Assume that an input X from station A needs to establish a connection to an output Y in station C. X may be a document produced by a minicomputer A and required to be printed as an output by a printer station C.

The input X at SAP A1, requests a connection to a station at SAP C1. The link layer at station A sends to the LAN a 'connection request' frame which includes the source address (A1), and the destination address (C1) which defines the SAP C1 where the printer is connected. Control bits are included to indicate that this is a connection request. The LAN delivers this frame to station C which, if it is free, returns a 'connected-accepted' frame to station A.

Henceforth, all data from X will be assembled into a series of frames by A's LLC, which includes source (A1) and destination (C1) addresses. Incoming frames addressed to (A1) will be rejected unless they are from (C1) (e.g. acknowledgements). Similarly, station C's printer is declared busy and C will only accept frames from (A1).

At the same time, a different input P, also from station A could attach to SAP A2 and exchange data with an output at SAP C2, and similarly a third

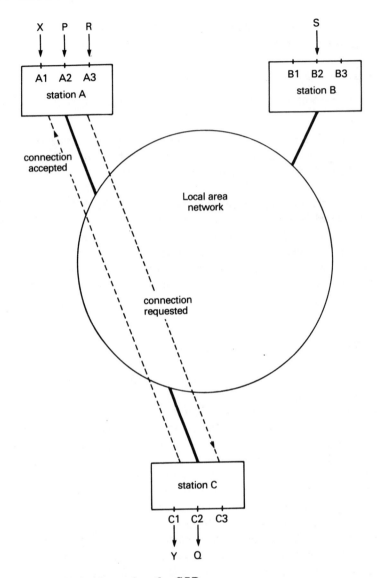

Fig. 8.8 Multiplexing using the SAP.

input R at SAP A3 could connect with an output S at SAP B2 etc. These are all examples of multiplexing a number of different input/ouput operations possible with the use of SAPs.

Two functions which are also contained in the LLC layer are:

a error control and

b flow control.

These are both end-to-end control processes arranged to provide acknowledgements and to ensure an error-free transmission across the network. The functions are provided in much the same way as with the supervisory frames in HDLC by the use of sequence numbers (NEXT and SEQ).

Finally, two additional functions are located in the LLC layer which are similar in operation to those found in ISO layer 3. These are:

a Datagram service: used for highly interactive traffic (this is an unacknowledged service).

b Virtual circuit service: to provide an acknowledged transport service.

Two special terms are associated with these last two functions. These are **connectionless** and **connection-oriented** and refer to services provided with these functions.

A **connectionless** service is that section of the protocol which supports the exchange of single data units (packets or datagrams) without the establishment of a data link connection between the link level entities. With a connectionless service there is **no acknowledgement mechanism** nor are there any flow control or error recovery mechanisms. This obviously makes for speed in the transfer of short messages, such as datagrams, but would not be satisfactory for the transmission of long messages where accuracy and notification of receipt are essential.

A **connection-oriented** service supports the fuller transmission of larger messages containing several packets. Here a data link connection is established between the two communicating link level entities prior to the exchange of level 2 data. Data units are delivered in sequence, error recovery functions are provided and flow control operates. At the end of data exchange link disconnection procedures are used to close down the link. The example of a service access protocol given in section 8.4.1 shows the operation of a connection-oriented service.

When we consider the application of these two services, we find that the main applications of interest for many users are terminal access, electronic mail and job transfer. All of these are connection-oriented in that they involve the establishment of a connection to a remote system, the reliable transfer of significant quantities of data and disconnection at the end of a transfer. Other applications such as data capture or monitoring functions would be considered as connectionless, since only a small amount of data is transferred and it may not be necessary for every data unit to be reliably received.

8.4.2 The medium access control layer

For a given LLC protocol, several different MAC options may be provided.

This is the protocol layer in which the differences in LAN topology are taken into account. For example the **bus/tree topology** requires a suitable protocol algorithm which arranges the suspension and retransmission of a data packet to avoid collision on the medium. The **ring topology** needs a mechanism to permit a data packet to be inserted into the ring at the correct time and to remove it when necessary.

The most commonly used medium access control technique for bus/tree topologies is CSMA/CD, considered in section 7.3.2[10]. This is one of a number of techniques referred to as **random access** or **contention** techniques. As mentioned earlier, the uncontrolled access to a common transmission medium such as a bus will result in numerous collisions, i.e. two stations transmitting at the same time resulting in severely corrupted data going along the bus. Some form of access control is needed.

CSMA/CD causes stations wishing to transmit to observe the rules already stated in section 7.3.2. Note that the last two rules, 4 and 5, form the CD part of CSMA/CD and are inserted so that when a station is transmitting a long frame (i.e. longer than the frame propagation time across the network) and a collision occurs, then this is detected and transmission ceases, to avoid further transmission of corrupted data. The implementation of these rules forms, of course, the MAC protocol for a bus/tree type of LAN. There is, however, a measure of imprecision in the mechanism, whereby a station involved in a collision operation decides to retransmit its message again. As originally conceived, the CSMA/CD protocol is sometimes called a **1-persistent protocol** where the 'persistence' refers to the mechanism for action if a competing transmission is sensed on the medium. Recalling that rule 2 of the protocol is defined as 'If the medium is busy, continue to listen until the channel is sensed quiet; then transmit immediately'. This is interpreted with 1-persistent protocol as the implementation of a **binary exponential back-off** technique, i.e. a station will attempt to transmit repeatedly in the face of repeated collisions, but after each collision the mean value of the random delay before the repeat transmission occurs is doubled. After 16 unsuccessful attempts the station abandons the message and reports an error. This system is found to be quite efficient over a wide variety of loading situations and a high network utilization is secured.

An alternative is not to transmit in the next free gap following collision sensing, but to wait a random amount of time before attempting retransmission. This is known as **non-persistent** CSMA and may be considered less selfish than the 1-persistent algorithm where a particular user may be able to exercise a longer control over the network than is considered reasonable by the other users.

A compromise is **_p_-persistent** where the parameter p determines the probability that the transmitter delays retransmission of its data. A new set of rules would be:

1 If the medium is quiescent transmit with a probability p ($p < 1$). This transmission is delayed by the length of one slot, also with a probability, this time of $(1 - p)$.

2 If the medium is busy, continue to listen until the channel is idle and repeat step **1**.

3 If transmission is delayed one time unit, repeat step **1**.

Although conceptually fairer, neither of these schemes (the non-persistent and p-persistent) are very widely used, since in a practical situation not enough collisions are actually experienced to justify the complication of a probability calculation and the non-persistent situation is wasteful of transmission time. As a consequence, although less than ideal, the 1-persistent or binary exponential back-off scheme is the one generally chosen for network implementation. A full account of these various methods of applying CSMA is given by Kleinrock and Tobagi to which the reader is directed[11].

Quite different MAC protocols are required for ring systems. For example, the Cambridge ring requires a MAC which can extract all the relevant control information from the circulating packet; noting from the full/empty bit in the control field if the packet can be used for transmission or if it contains circulating information. If the latter is the case it needs also to read the source and destination addresses in their respective fields and if the destination address corresponds to that of the station reached by the circulating packet, then the data will need to be read into the station also. This means a considerable overhead in control bits is required for each packet. Additionally, a station can only send one packet per round trip of the ring. The bandwidth efficiency is not, therefore, very great but the system is much simpler to operate and more reliable than CSMA/CD. The packet format was shown in Fig. 7.4b and corresponds to the contents of one slot circulating around the ring past all the connected station. Each slot contains:

a **Two 8-bit addresses**: for source and destination of the data;

b **Two 8-bit data bytes**; and

c **five control bits**.

Two of these control bits are set by the addressed station to inform the sender whether it has accepted the data or is busy. The full slot makes a complete round trip, to be marked empty again by the sender unless this has further data to send when it would make use of this empty slot the append further data. The now-empty slot continues around the ring until it reaches a station with data to send which will proceed to convert the control bit to indicate 'packet full' and append the data and address information.

The Cambridge ring is supported by a protocol developed for JNT academic use and known as the CR-82 ('orange book') protocol[12]. Current commercial standards adhere to CR-82 but generally apply to a 40-bit frame.

CR-82 is, essentially, a virtual circuit protocol and, unlike the original Cambridge ring operation, does not support a datagram approach. This is appropriate to linking a network to a WAN, such as JANET or PSS, but the short packets involved do lead to inefficiencies when not handled as a series of unacknowledged datagrams. Further, in a common practical case of single station operation, only one minipacket can be sent at a time, other packets on the ring remain empty for the duration of the transmission, and this also leads to functional inefficiencies. In its favour, however, the access procedures for the Cambridge ring are relatively simple when compared with other ring procedures (e.g. token ring) and its highly predictable behaviour under conditions of heavy loading makes it easy to manage.

The MAC protocol required for a token ring system permits a larger amount of data to circulate around the ring when this is in the active state. The technique is based on the use of a single token that circulates around the ring when all stations are idle. Referring to Fig. 7.7, a typical example of a token is an 8-bit pattern 01111111. A station wishing to transmit must wait until it detects a token passing by. It then changes the token from a 'free token' to 'busy token' by changing the last bit in the token (e.g. to 01111110). Note that since two specific bit patterns are defined for the free and busy tokens, these must not appear anywhere in the address or data. To avoid this a process of 'bit stuffing' is applied, as described previously in section 4.3.1. With the free token changed to a busy token the station proceeds to transmit a frame immediately following busy token.

Since there is now no free token on the ring, other stations must wait. The frame on the ring makes a round trip to meet the transmitting station again which erases its data. The transmitting station will then insert a new free token on the ring. When a transmitting station releases a new free token, the next station along the ring (remember that packets only circulate in *one* direction) will be able to seize the token and transmit. Thus the use of a token guarantees that only one station at a time may transmit.

In principle, once the token is held, very long packets can be transmitted. It is conceptually possible for a given node to hold the token for as long as it likes whilst transmitting a lengthy message. In practice, a limit is set on how long a token may be held so that a 'lost token' failure can be quickly detected. In some token rings a node may only be able to transmit one packet while it holds the token; in others several packets may be sent in sequence, providing a given time limit is not exceeded.

8.5 IEEE 802 LAN STANDARD PROTOCOLS

A more formalized application of the LLC and MAC concepts described above has been attempted through the definition of a set of LAN standard protocols which meet a range of network requirements for a wide variety of users. These are the protocols developed under a cooperative project, known as IEEE Project 802. The project originated at a gathering of interested professionals at San Francisco in the early 1980s[13,14]. A considerable amount of work was put into the development of a LAN standard which would be universally recognized by the computing community. Many organizations and individuals from a number of countries participated. These included industrial representatives from the USA and Europe, standards organizations, such as the National Bureau of Standards and the British Standards Institute, telephone service companies, e.g. Bell Labs, Comsat and many others. In a remarkably short space of time a set of procedures was agreed and an initial set of draft standards published. These have since been formalized and are now widely accepted, together with the ISO seven-layer OSI standards, with which they are compatible.

The IEEE LAN Project 802 restricts its standardization activities to the two lower layers of the ISO reference model, namely the data link layer and the physical layer. The ground rules for design were:

a HDLC convergence,

b transparency to topology,

c transparency to transmission rate, and

d transparency to media.

HDLC convergence means that the specifications adhere as closely as possible to the HDLC standard, which already commands widespread allegiance in the networking community. Transparency can be interpreted as 'independent of' in respect of the operation of higher layers in the protocol.

To achieve these goals the normal data link control and physical layers are repartitioned as shown in Fig. 8.9. In the IEEE 802 standard the MAC layer has several options dependent on the network topology. Current standards are:

a IEEE 802.3 CSMA/CD bus network,

b IEEE 802.4 Token-passing bus network,

c IEEE 802.5 Token-passing ring network, and

d IEEE 802.6 Metropolitan area network.

This last has yet to be defined fully and released by the IEEE 802 Committee. There are two specifications which preface the four LAN descriptions:

a **Specification IEEE 802.1** which is a guideline document explaining the relationships between the 802 standards and the ISO OSI reference model.

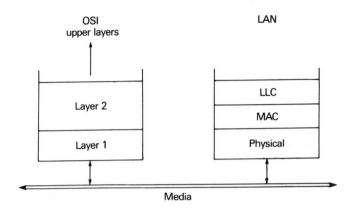

Fig. 8.9 IEEE 802 protocol layers.

    b **Specification IEEE 802.2** which describes a uniform interface between user equipment and four different cable technologies.

Detailed descriptions of these standards are available in refs. [15-18].

    The requirements for the IEEE 802 LAN frame are very similar to HDLC, namely:

    a A data or information field.

    b A control field.

    c Starting and ending patterns.

    d Address fields.

Frame formats and address conventions are common to all the types and levels of the IEEE 802 protocols.

    The address fields provide a major difference with HDLC. Because LAN links are multiple-source, multiple-destination, both source and destination addresses are required. Further, unlike HDLC and virtually all other layer 2 protocols, the IEEE 802 LAN supports a form of multiplexing common in layer 3 protocols. It does this by making use of the SAP as discussed earlier in this chapter.

    The interconnection implementations for the 802 protocols are designed for data rates in the range of 1–20 Mbps in 5 Mbps increments. Alternative data rates are supported for some media as we shall see later.

    One of the most interesting aspects of IEEE 802 is that it is not attempting to choose the 'best' protocol. Instead, it is defining a layered approach where, at the lowest layer, a network owner may choose any one of a number of cable technologies satisfying his cost/performance/security requirements.

    The complexity and bulk of control procedures are contained in the common LLC layer protocol and the various protocol options for the MAC layer. As before, we need to consider their operation separately.

8.5.1 LLC standard 802.2

This standard defines the function and operation of the logical link control sublayer which constitutes the top sublayer in the data link layer shown in Fig. 8.9. The LLC sublayer is common to the various media access control methods (MAC) which are discussed in the next sections.

In accordance with the OSI multilayer protocol methodology, the frame structure becomes more complex as the message permeates down through the LLC, MAC and physical layers. This is indicated in Fig. 8.10a. To the message reaching the LLC layer from the transmission node (or upper protocol layers if they exist) will be added a header containing a destination service access pointer (DSAP), a source service access pointer (SSAP) (both of which refer to the LLC procedure for the data being transmitted) and a control (C) byte. The MAC layer will add an access control field (AC), a destination address (DA), and a source address (SA) to the trailer accompanying the message, together with a frame check sequence (FCS). Finally, the physical layer will add a start frame delimiter (SFD) to the beginning of the frame and an end of frame delimiter (EFD) at the end. The complete frame structure is shown in Fig. 8.10b.

As an indication of the way in which these protocol standards function we can look at the interface between the LLC and MAC sublayers. This may be considered as a logical area in which the logical exchange of information and control messages between the two layers takes place. The exchange operates through primitives which we discussed earlier in section 8.1.1.

These are:

a **LLC primitives** used across the interface between the upper layers and the LLC sublayer.

b **MAC primitives** used across the interface between the LLC and MAC sublayers.

There are also **LLC protocol data units** which are exchanged between LLC entities in corresponding (peer) sublayers in the communicating nodes (see Fig. 8.2).

The primitives used in all the 802 standards are of three generic types:

a Request: This is passed from layer N to layer $N-1$ to request that a service be initiated.

b Confirm: This is passed from layer $N-1$ to layer N to convey the results of the previous request primitive.

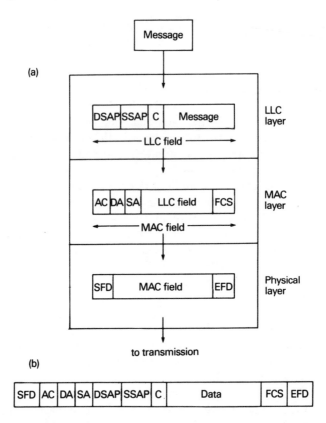

Fig. 8.10 (a) Communication with the IEEE LLC 802.2 standard.
(b) Frame structure.

  c Indication: This is passed from layer $N-1$ to layer N to indicate an event
                     is taking place in the $N-1$ layer that is significant to layer N.

Various combinations of these are used across the sublayer interfaces.
Examples are:

  Data request: Connection request

  Data confirm: Disconnect indication

Semantics for these follow the form:

L.DATA.REQUEST (local address, remote address, link layer service data
unit, service class).

  These commands and responses are not numbered, and to maintain integrity the LLC starts a timer each time it transmits a command and retransmits the command if no response has been received before the time-out period has elapsed. After a number of retransmissions have been carried out, the LLC will abandon its attempts and inform the sender that the station requested cannot be contacted.

Messages are exchanged between entities in the corresponding peer sub-layers through the use of **protocol data units**. These are known as **peer protocols** and use the services of the underlying layer to effect the successful transfer of information from one location to another. The structure of the protocol data units is based directly on HDLC frames. There are some important differences, however. An extended address and control field width allows both source and destination address to be specified and a group destination address if required. Some additional commands are provided but the use of flags, abort sequences and bit stuffing are excluded.

To satisfy a range of potential applications, two types of data link control operations are included. In the example given above a **connection-oriented** service was implied. The other service is a minimal **connectionless service** providing a datagram service with no acknowledgement. Both were described earlier in section 8.4.1. In the 802 terminology they are referred to as Type 2 and Type 1 data link control, respectively.

The alternative MAC sublayers differ in the media and topology that they can employ. Both broadband and baseband techniques are covered. In the following sections, each IEEE 802 MAC standard will be considered in terms of the current release up to 1984. It is convenient to consider the operation of each MAC standard separately together with the physical layer, which is essentially layer 1 of the OSI standard. Note that, in addition to being different for each media access type, the physical layer is also further sub-divided within each type of medium to cover the application of twisted pair, coaxial cable (in some cases several alternative coaxial cables), fibre optics etc.

8.5.2 CSMA/CD standard 802.3

The 802.3 procedure is logically the same as that used in Ethernet. The procedure permits all stations on the bus to contend for access within a given time slot equal to the round-trip propagation delay for the full length of the bus. A given station will monitor traffic on the bus. When none is detected within this time slot, the station may load one addressed data frame on to the bus and transmission commences. Of course, two stations may do this at the same time and this will cause contention to occur. In this event both stations will back off and try again at a later time as described previously. The 802.3 CSMA/CD procedure differs from the Ethernet definition in certain areas, principally in frame structure and physical interface so that the two definitions are not exactly compatible, although almost identical in operation.

As with any other layer organized protocol, we need to consider interfaces on both sides of the layer. With the MAC layer these are:

1 The interface between the MAC sublayer and the LLC sublayer. This includes facilities for transmitting and receiving frames and operating status information for use by the higher-layer error recovery procedures.

2 The interface between the MAC sublayer and the physical layer. This includes signals from framing (carrier sense, transmit initiation), and contention resolution (collision detect), facilities for passing separate bit streams (transmit, receive) and a wait function for timing.

The process of transferring information (data and control) relies on the application of a range of service primitives developed from the three generic types mentioned earlier. For example, the process of conveying data between the MAC and LLC layer would require the use of three primitives:

a   MA-DATA.request,

b   MA-DATA.confirm,

c   MA-DATA.indication.

A request from a higher protocol layer for data would result in the generation of an MA-DATA.request which would include a set of parameters as follows:

MA-DATA.request (destination address, m-sdu, service class)

The destination address parameter would specify the address in the LLC entity where the data are to be placed. The m-sdu parameter specifies the MAC service data unit to be transmitted and enables the length of the data unit transferred to be specified. The service class parameter is inoperative at present since only one class of service is available with IEEE 802.3. The effect of this request would be to cause the MAC sublayer to append the destination address and the source address (already included in the transferred data) to the data and to locate the properly formed frame for transmission upon demand. Response to this request would result in the generation of a MA-DATA.confirm primitive to provide an appropriate response for the request. The appended parameter (transmission status) would serve to indicate the success or failure (e.g. a collision) of the request.

Finally, following the actual transfer of data, the MA-DATA.indication is invoked by the MAC sublayer entity for transmission to the LLC sublayer entity. The semantics are:

MA-DATA.indication (destination address, source address, m-sdu, reception status)

This primitive would be generated only if a properly formed frame is received correctly at its destination. Similar sets of operations are carried out to control transfer of data and control information across the MAC/physical layer interface.

The format of the frame transmitted on the network is shown in Fig. 8.11 and will be seen to be very similar to the Ethernet frame format. It contains:

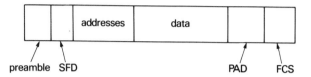

Fig. 8.11   IEEE 802.3 MAC frame structure.

a   **Preamble field**: This is a seven-octet field used to allow the physical layer circuit time to reach its steady state synchronization with the received frame timing.

b   **Start frame delimiter (SFD) field**: This is a sequence, 10101011, which indicates the start of a valid frame.

c   **Address fields**: The two fields are destination address and source address and may be either 16 or 48 bits long. (More than one destination address may be implied if a group or broadcast address is being used.)

d   **Length field**: This is a two-octet field which indicates the number of data octets in the following data field.

e   **Data and PAD fields**: A variable length of n octets is used for data. A minimum frame size is needed for correct operation and, if the length of data is smaller than this, a number of PAD (padding) octets are added.

f   **Frame check sequence (FCS) field**: The computation of this field, used for error checking with a cyclic redundancy checking procedure, covers all the above fields except for the preamble, SFD and FCS. The **encoding sequence** (see section 4.4.2) is defined as a polynomial:

$$p(x) = x^{32} + x^{26} + x^{23} + x^{22} + x^{16} + x^{12} + x^{11} + x^{10} + x^8 + x^7 + x^5 + x^4 + x^2 + x + 1$$

It is useful to look at the operation of the IEEE 802 procedure in terms of the transmission of a frame between sublayers. We can consider first transmission and reception without contention and then look at what happens when there is access interference (i.e. a collision) with subsequent recovery to normal transmission.

During normal operation in the **transmission phase** the following procedure takes place: when an LLC sublayer requests the transmission of a data frame across the medium a process of frame construction begins. A preamble and SFD is appended to the beginning of the frame. A PAD is appended, if necessary, to ensure that the transmitted frame length reaches a minimum frame size. Also appended are the destination and source addresses, a length count field and an FCS for error detection. The modified frame is then passed to a media access management component in the MAC sublayer for transmission. This management process monitors the network traffic and, when the medium is clear, a serial bit stream is passed through the physical layer interface for transmission. When transmission has been completed without contention, the MAC sublayer informs the LLC sublayer and awaits the next request for a frame transmission.

During the **reception phase** the arrival of a frame is detected at the physical layer which responds by synchronizing with the initial preamble data sequence and initiates the carrier sense signal. Subsequent bits in the frame are passed up to the MAC sublayer, where the leading bits are discarded up to and including the end of the preamble and the SFD. In the MAC sublayer the frame's destination address field is checked to determine whether the frame should be received by the station. If this is the case, then the truncated frame is passed to the LLC sublayer having been checked for invalid MAC frames through the application of the FCS.

When access interference occurs, such as during **collision operation**, the physical layer will detect the malfunction and initiate a collision detect signal. This is received by the media access management component of the MAC sublayer and the collision procedure is commenced. First, the management enforces the collision by transmitting a jamming bit sequence to all other stations on the network. This ensures that the collision event is noted by all stations, which will individually initiate their own waiting period if they also have a message to transmit. After a random selected time period, the station attempts to retransmit. It may have to do this several times. The management section monitors these attempts and adjusts the waiting period before retransmission by applying a backing-off algorithm as discussed earlier. Eventually, either the transmission succeeds or is abandoned on the assumption that the medium has failed. Note that the bits resulting from transmission collisions are received and decoded by the physical layer just as a valid message would be. It is one of the tasks of the MAC sublayer media management component to distinguish these from valid transmissions.

8.5.3 Token passing bus standard 802.4

The token bus uses a procedure of passing a token which confers the right to transmit between stations connected to the bus. When a station, having use of the token, finishes its transmission, it must pass the token to another station.

Each station maintains a pre-defined table which it uses to locate the next station in the logical transmission list (not necessarily a physically adjacent station) to receive the token if it does not itself wish to transmit. In effect, the table transforms the bus into a **logical ring**. Hence, in the operation of a token bus we have the situation of a **broadcast medium**, but due to the addressing procedures detailing where next to pass on the token, the bus acts as a **sequential medium**. A problem with the token bus is the failure of a station to pass on the token either through contention or faulty operation. The house-keeping functions needed to recover from the lost-token situation increase the complexity of the token bus media access method.

The medium access method used is applicable to modem operation where three different forms of signalling on 75 ohm coaxial cable are considered:

a   Phase-continuous FSK at a data rate of 1 Mbps using Manchester coding with:

f_0 = 6.25 MHz (digital 0)
f_1 = 3.75 MHz (digital 1).

b   Phase-coherent FSK at two possible data rates, 5 and 10 Mbps with:

f_1 = 5 MHz (at 5 Mbps) f_1 = 10 MHz (at 10 Mbps)
f_0 = 10 MHz (at 5 Mbps) f_0 = 20 MHz (at 10 Mbps).

c   Multi-level duo-binary coding (QAM) at three possible data rates, 1, 5 and 10 Mbps and using channels in the range 59.75–264 MHz.

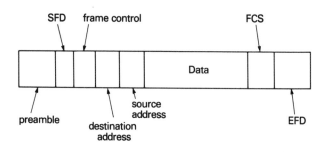

Fig. 8.12   IEEE 802.4 MAC frame structure.

The MAC frame format is shown in Fig. 8.12 and the fields used are:

a   **Preamble field**: This is a pattern to set the receivers' modem clock and level.

b   **Start frame delimiter field**: This is a unique 8-bit pattern having a similar function to the SFD in the CSMA/CD frame.

   c **Frame control field**: This determines the class of frame being transmitted, i.e. MAC control, LLC data or general management data.

   d **Destination and source address fields**: Again either 16 or 48 bits as with CSMA/CD.

   e **Data field**: This also contains the LLC protocol data unit and a MAC management frame but no padding field since there is no minimum frame length.

   f **Frame check sequence field**: This covers the previous fields but not the following EFD.

   g **End frame delimiter (EFD) field**: This indicates if more than one frame is to be sent. Another bit, the (E) bit, indicates that an error has been detected in the frame which is nevertheless transmitted.

The standard also includes a specification of a broadband bus system designed to operate in a bi-directional split frequency manner as described in Chapter 7.

8.5.4 Token-passing ring standard 802.5

Here the token is passed around a ring without having to address the stations. This is because the ring behaves as an actual sequential operating ring rather than as the logical operating ring as realized in the token bus. A station wishing to transmit data needs only to capture the token as it passes by and to release it after the data transfer. Stations with no transmission requirements simply ignore the passing token.

A simple station address sorting algorithm is maintained to determine the initial station order in passing the token around the ring. This algorithm is resorted to if the token becomes lost or corrupted during operation.

Implementation of IEEE 802.5 follows the general scheme already described in section 8.3.2. The key elements are:

   a Single-token protocol: A station that has completed transmission will not issue a new token until the busy token returns.

   b Priority bits: These indicate the priority of a token and hence which stations can use the token next. It may be set by station or by message.

   c Monitor bit: Used if a central ring monitor is incorporated.

   d Reservation indicators: These are used to allow stations with high priority messages to indicate in the frame that the next token is to be issued at the requested priority.

e Token-holding timer: Started at the beginning of data transfer, it controls the length of time a station may occupy the medium before transmitting a token.

f Acknowledgement bits: These are set to 1 for the following:

error detected (E),

address recognised (A),

frame copied (C).

They are reset to 0 by the transmitting station.
Any station may set the E bit. Addressed stations may set the A and C bits.

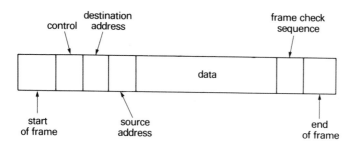

Fig. 8.13   IEEE 802.5 MAC frame structure.

The frame format is somewhat less complex than the token bus since modem control is not required. This is shown in Fig. 8.13 and the fields used are:-

a **Starting delimiter field:** A unique 8-bit pattern as in CSMA/CD.

b **Access control field:** This has the format 'PPPTMRRR', where PPP and RRR are 3-bit priority and reservation variables, M is a monitor bit, and T indicates whether this is a token or a data frame.

c **Frame control field:** This indicates whether this is an LLC data frame. If not, bits in this field control operation of the token ring MAC protocol. This is also the field where priority operation is indicated.

d **Destination and source address fields:** Again these can be either 16 or 48 bits long.

e **Data field:** No maximum length — it can actually be of zero length.

f **End frame delimiter field:** As in token-passing bus.

g **Frame check sequence field**: As in token-passing bus.

h **Frame status field**: This contains the address recognized (A) and frame copied (C) bits which are set by the receiving node.

A combination of the **frame status** and **end frame delimiter** allows the status of the station to be checked.

In operation, stations in the received mode 'listen' to the signals on the ring. Each station can check passing frames for errors and set the E bit if an error is detected. If a station detects its own address it sets the A bit to 1; it may also copy the frame, setting the C bit to 1. This allows the originating station to differentiate three conditions:

a station non-existent/non-active,

b station exists but frame not copied,

c frame copied.

**SUMMARY
Chapter 8**

The difficulties in devising a suitable and comprehensive protocol to permit information to be exchanged freely between many different types of equipment operating at a range of speeds over different media has led to the development of a **layered model protocol**. Here the various tasks are structured to be carried out by different sections of the protocol. An agreed international protocol is the CCITT seven-layer model. The upper three layers of this model are concerned with exchange of application data and files and include exchange codes to permit inter-machine connection. These layers, together with the remaining layers, are applied in a wide area network containing many widely distributed nodes and connected host computers.

The lower three layers are specifically designed for the exchange of data, including multiplexing of data streams, data checking and acceptance of alternative transmission media. These are the communication layers which form the basis of local area network interconnection.

The requirements for a LAN protocol are less stringent than in a WAN, since a complex mesh network is not used and end service requirements are simpler. Many of these protocols are based on a high-level data system link (HDLC) procedure which is concerned with transmission of packets or frames between nodes on a network.

Developed from this basis are a series of IEEE 802 standard protocols. These consist of a common logical link control (LLC) protocol and a number of alternative medium access control (MAC) protocols to permit control of bus or ring networks and the use of different media.

Operational requirements for these two types of network are quite different. In the case of a **broadcast** network, such as a bus, there is a need to allow for contention between stations competing for network facilities. A protocol developed for this purpose is called CSMA/CD. For a **sequential** network, such as the slotted or token ring, contention can be avoided but it is necessary to prevent continuous circulation of data around the ring and to arrange suitable mechanisms for data insertion.

Finally, in order to accept or provide data for packet transmission, a packet assembler/disassembler device (PAD) is required, either as part of the node or ancillary to it, which serves to partition the data into short packets of equal length.

**PROBLEMS
Chapter 8**

P8.1 List the seven layers of the CCITT ISO architecture for network communications. **a** Describe their function and justify the existance of each one. **b** Which layers are essential to LAN communications and why?

P8.2 Describe in detail, using an explanatory diagram, the sequence of events by which DTE 'A' may request a virtual connection to another DTE 'B' connected to the same network. Then describe how data are transferred in packets concluding with the clearing of the temporary connection from the DTE 'B'.

P8.3. Assuming HDLC protocol, **a** Distinguish between the normal response mode and the asynchronous mode of working. How are they defined in the HDLC frame structure? **b** How is flow control achieved through this frame structure?

P8.4 Describe the function of the logical link control and medium access control protocol layers as defined in the IEEE 802 standards and indicate their relationship with the lower protocol layers in the ISO seven-layer reference model.

P8.5 Consider a satellite communication system using the HDLC protocol for communication. The transmission rate is 1 Mbps and the frame length is 1024 bits. Given that the propagation delay for a complete up/down transmission is 270 ms, what is the maximum possible data throughput?

P8.6 Two fields in the HDLC information frame are called SEQ and NEXT. How may these be used in the control of a sliding window flow control system?

9

CONVERGENCE OF DIGITAL SERVICES

Until quite recently, the various forms of electronic communications (speech, text, vision, data processing, office technology, distance control etc.) have developed quite separately with the major changes occurring in the transition from analogue to digital technology.

Now that digital methods are available for each separate communication requirement, a position has been reached when it becomes possible to consider a convergence of these facilities in order to integrate them into one common digital transmission service. The combined service is called an **Integrated Services Digital Network** (ISDN) and a number of countries are actively engaged in the development and implementation of such a service with mutual adherence to a set of shared international standards. Apart from the convenience and economy of scale that is being achieved, this process will allow the development of new ways of using communication media not feasible with separate communication channels.

One of these new facilities is the incorporation of **message handling systems** (MHS) and **message transfer systems** (MTS) operating under national and international auspices. Unlike the present telex and teletext services, these newer facilities are designed to operate on a person-to-person basis rather than between machines. International standards have been agreed for these facilities which permit full integration to communication systems already in use and operating under OSI protocols.

On a quite different level, convergence of techniques is also occurring for industrial usage. It concerns the way in which an industrial organization coordinates its various data processing and machine control activities, covering such operations as stock control, financial accounting, production control, manufacturing processes and office procedures. The two techniques that have arisen to serve this purpose are known as **manufacturing**

applications protocol (MAP) and **technical and office protocol** (TOP). These are also designed to exploit the seven-layer architectural protocols described by OSI.

Future areas for convergence lie in the area of printed information production and distribution (books, newspapers, journals etc.) and in their coordination with existing network-accessible information services (data bases). Standards for these have yet to reach the stage of international acceptance, although a limited number of software techniques for text editing and printing are widely accepted in the printing industry.

In this chapter, attention will be directed primarily to those areas where widespread agreement for common standards is evident, notably ISDN, MHS, MAP and TOP, in order to provide an introduction to present day activities in convergent techniques.

9.1 INTEGRATED SERVICES DIGITAL NETWORK

9.1.1 Early developments

Prior to the development of separate telex and data networks, digital or non-speech services were provided through the public switched telephone network (PSTN). This resulted in a limited **integrated services analogue network**, supporting only a few such services. The design of a completely digital transmission service, the **integrated digital network** (IDN) was made possible by the introduction of digital transmission using PCM and the use of digital switching techniques with TDM replacing FDM applied in the analogue PSTN[1].

IDN development is seen most clearly where a communication system is contained within a small area, such as a company or other self-contained organization. An early development was the evolution of the **private automatic branch exchange** (PABX) into the fully digital equivalent **computerized branch exchange** (CBX) which represents the modern on-site telephone communications system for a small company (Fig. 9.1).

The CBX offers a very extensive interconnect capability with outlets throughout an organization. However, many PABXs use old technologies with high connection establishment and release time. They are also designed for short duration speech calls and the high-density traffic patterns required for data demanded elaborate flow control procedures, particularly since the process is limited to traditional speech circuit transmission speeds.

The installation of **stored program controlled exchanges** overcame a number of these difficulties through the use of fully digital systems in which speech is carried as a 64 kbps PCM signal. The digital capability can then be extended to provide a high-speed mixed speech and data transmission system

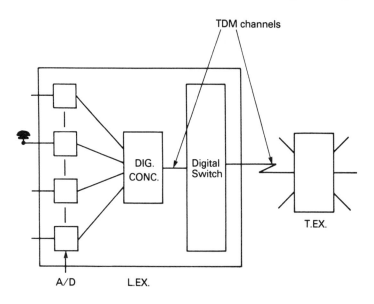

Fig. 9.1 A computerized branch exchange.

at the telephone terminations. A modern CBX will have line interfaces which multiplex up to four 64 kbps duplex digital channels over normal telephone extension wiring. The new generation of CBXs also overcomes other major deficiencies of the old PABX technology. Call establishment and clearance times are shorter and data traffic does not block the normal speech switching capacity of the exchange. In addition, the CBX can utilize the spare capacity of the control processing system to provide a message handling facility or **mailbox** and its analogue equivalent, the **speech mailbox service**, where the (digitized) speech message is held in a digital store. These services may be provided to any CBX line and extended over analogue or digital trunk lines to other CBXs or PABXs.

The integration of the digital CBX to a digital network carrying data communications and shared facilities within the organization (such as word processors, telex, document transmission, common storage etc.) can extend the value of the CBX still further[2]. Where the CBX is connected to a digital network in this way it can also extend the connection to a higher level of digital multiplexing within the CBX, typically a 2 Mbps trunk service which would normally serve a group of 32 external PCM connections. Access from a terminal to the network is obtained by normal call routing within the CBX.

Clearly, even if the interchange standards are common throughout the community, access from a CBX-based system terminal to other parts of a network will be restrained by the maximum speed of the CBX digital channels so that, for example, the movement of whole files of information will be less rapid than in a connected higher-speed network. However, the combined system can be used in many organizational structures, and the data service extended to include distributed processing elements, computer mainframes, file storage, mailbox and links to various networks services.

9.1.2 The ISDN concept

These ideas have now been carried further with the ultimate ambition to link all forms of data communications together through the national digital network facilities, the PSDN, and, where applicable, to certain private networks. The main direction of this development is defined in the work of the standards organizations and particularly the CCITT who provide the major recommendations to be agreed internationality.

ISDN has been defined by the CCITT as 'a network which provides direct digital connection between users in order to support a range of different telecommunications services.'[3] The principal features of the CCITT recommendations are:

a   Support of a wide range of speech and non-speech applications in the same network.

b   Provision of a wholly digital connection between users' connected to the ISDN.

c   Provision of a range of services using a limited set of connection types and multi-purpose interfaces.

d   Support of both switched and non-switched connections.

e   Provision of both circuit- and packet-switched connections and their concatenation.

f   Provision of a capability within the ISDN to permit the control of various service features, maintenance and network management functions.

g   Compatibility with a layered set of protocols defined by the ISO seven-layer standards.

Essentially, the evolution from IDN to ISDN is the conversion of the users' access connection from the present analogue system to a completely digital transmission path. This allows support for both speech and non-speech services on the same digital network directly from the users' premises.

9.1.3 International standards

A set of recommendations for ISDN were agreed in 1984 through the CCITT in Europe and are referred to as the I-series recommendations[4]. It is to these recommendations that current systems are being designed, the most important of these being:

a I-120 Integrated services digital networks.

b I-210 Principles of telecommunications services supported by an ISDN.

c I-310 ISDN — Network functional principles.

d I-340 ISDN — Connection types.

e I-411 ISDN user network interface — Reference configuration.

f I-412 ISDN user network interfaces — Channel structure and access capabilities.

g I-420 ISDN full user network inteface.

h I-430/440/450 Protocol layer specifications.

9.1.4 Network Structure

An architectural model of an ISDN is shown in Fig. 9.2. Since the ISDN has evolved from the PSDN, the main building blocks are a number of 64 kbps transmission channels each suitable for supporting a single speech channel and a common channel used for out-band signalling. However, as the ISDN is designed to support services requiring a higher bit rate connection, such as video and holography, some channels are also required at a higher bit rate. These operate at a rate which is a multiple of 64 kbps, typically 2.048 Mbps.

The choice of bit rate for the end digital connection is a compromise between the multimode operations required and the capacity of existing feeder lines. Some of the criteria guiding this selection are as follows:

1 The need to support PCM-encoded speech at 64 kbps.

2 Requirement for signalling capability. Since multimode channels are to the supported there are significant advantages in choosing out-band signalling for this.

3 Support of more than one simultaneous channel to different end stations. Thus a minimum requirement for a PCM-encoded signal would be a further 64 kbps channel.

4 Line feeder and equipment required for end user access must be capable of carrying the data rate required without the inclusion of additional repeater equipment.

5 Power feeding of the user's equipment must be possible where the PSTN is involved, without an increase in exchange supply potential.

Although there is some conflict between these requirements it is now

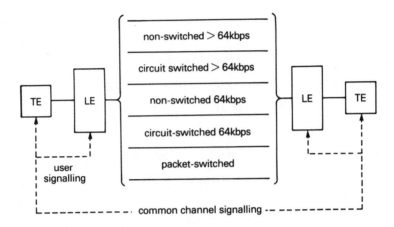

Fig. 9.2   Architectural model of an ISDN.

generally agreed that the CCITT I420 recommendation will provide a satisfactory basic access using a composite rate of 144 kbps. This rate is subdivided into:

a   a 64 kbps data channel (the B channel),

b   a 64 kbps data channel (the B′ channel), and

c   a 16 kbps signalling channel (the D channel).

The **B channels** are the basic user access channels and can carry any one of the following types of traffic:

a   PCM-encoded digital speech,

b   digital data for circuit-switched or packet-switched applications,

c   a mixture of lower-rate traffic, including digital data and digitized speech encoded at a fraction of 64 kbps.

The **D channel** serves two purposes. First, it is used to exchange control information between user and network (**out-band signalling**). Second, it supports lower-speed   digital data requirements. If the D channel were restricted solely to signals associated with the control of the calls over the B channels it would only be used for a very small fraction of the time. However, since fast call setup and clearing are required for the B channels, the D channel must be available for signalling when required. For this reason the D channel is only suited to carry low-speed packet data in addition to signalling, the signalling (itself packeted into messages) being interleaved with the packet data and having priority over packet data transmission. There are a few further restrictions in the use of the D channel for data transmission, apart from its lower priority, such as a lower data rate (16 kbs) and a maximum defined packet length.

Note that the ISDN gives to the user three connections to the local exchange (2B + D) using the same copper wire pair that originally provided a duplex analogue telephony service. In addition, the system provides a capability of a 64 kbps switched data service which has not previously been supported over the PSTN using modem conversion.

The type of data traffic supported by the B and D channels is summarized in Table 9.1. These channel types are grouped into transmission structures that are offered as a package to the user. They are the **basic rate access system** and the **primary rate access system**.

Table 9.1

Data traffic supported under ISDN

B channel (64 kbps)	D channel (16 kbps)
Digital speech	Signalling
64 kbps PCM	Basic
Low bit rate (32 kbps)	Enhanced
High-speed data	Low-speed data
Circuit-switched	Videotext
Packet-switched	Teletext
Other	Telemetry
Facsimile	Surveillance
Slow-scan TV	Control

The basic rate access system is the CCITT 144 kbps system described earlier and is designed for domestic and small business needs.

The primary rate access system is at the next higher level of multiplexing and is designed for large business requirements where the PABX or CBX needs many connections to the local exchange. In Europe, the rate is 2.048 Mbps, consistent with that employed by the PSDN giving 30 B channels plus a D channel operating at 64 kbps and one unallocated channel. In North America, the standards are slightly different at 1.544 Mbps offering 23 B channels and a 64 kbps D channel. Connections at other rates have been defined by CCITT as[5]: H_0 channel operating at 384 kbps, H_1 channels — H_{11} operating at 1535 kbps and H_{12} operating at 1920 kbps, but their use has yet to be agreed internationally.

9.1.5 Network interface

The types of configurations possible for the ISDN user network interface are shown in Fig. 9.3. This illustrates two important concepts used with ISDN interface design namely:

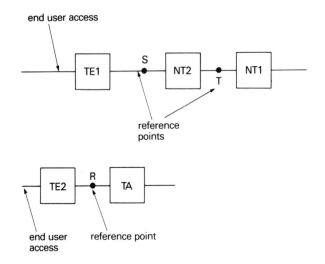

Fig. 9.3 User network interface configurations.

a **Functional groups**. Certain finite arrangements of physical equipment or combinations of equipments used to carry the user access.

b **Reference points**. Conceptual points used to separate groups of functions.

The points labelled R, S and T in Fig. 9.3 are reference points at which a physical interface may occur. The S and T reference points allow parallel connection to a number of terminals by employing the concept of a **passive bus**. Two versions of this bus have been defined: a short bus of maximum length 100 m, and an extended bus for a cluster of terminals up to 500 m from the appropriate functional box.

Reference points represent a useful way of identifying the point at which one function (generally a protocol layer) hands over to another function (layer). Note that this does not necessarily mean that the reference points are actual physical interfaces, since two or more functional groupings may be combined together in a single piece of equipment.

These functional groups are:

a **NT1 (Network terminator 1)**. This provides functions broadly equivalent to layer 1 of the OSI model, e.g. multiplexing, electrical transmission, timing, and power feeding of the 2B + D channels.

b **NT2 (Network terminator 2)**. This provides functions relating to the distribution of access to the user's network connections, e.g. PABX or CBX, local networks or work station clusters, and includes protocol handling, multiplexing and switching operations.

c **TE (Terminal equipment)**. Two types of TE have been defined. TE1 is the terminal equipment which is associated with an ISDN user network interface. TE2 is a terminal equipment which is associated with a non-ISDN defined data or an analogue network such as a CCITT X or V series interface.

d **TA (Terminal adaptor)**. This carries out the functions necessary to adapt the user interface of a TE2 to that of the CCITT ISDN user network interface.

In the user network interface shown in Fig. 9.3, the transmission line conveys the digital data to NT1 where they are modified to include the functional protocol information and to present this at the reference point T. The functional box NT2 then provides user access at its reference point S. In its most commonly used form NT2 could be an eight-way parallel highway extending the T reference point to eight S reference points for connection to user equipment (e.g. telephone or work stations). The I420 interface applied here refers to an eight wire parallel connection in which four wires are required for the duplex connection and the remaining four wires for power feeding.

Protocols designed for use in the ISDN are implemented through the OSI seven-layer model and many of the features of existing interfaces, e.g. the X25 interface, may be recognized in the following description of the ISDN basic and primary rate access network interface.

9.1.6 Protocol layer specifications

The **basic rate access network** conforms to recommendation I.420. Both this and an associated series of recommendations encompass circuit-and packet-switched access. The same basic procedures are used for network access, with a separation into different types of call taking place at the exchange.

The I420 recommendation refers to three other layers defined separately as:

a layer 1 recommendation, I430,

b layer 2 recommendation, I440 and I441,

c layer 3 recommendation, I450 and I451.

Layers 2 and 3 are only specified for the D channel and at present layer 3 recommendations refer only to circuit- and packet-switched control signalling. Layer 1 recommendation, I430, is equivalent to the physical layer of OSI and is an eight-wire interface permitting up to eight terminals to be supported on a passive bus via the NT2. This layer is somewhat more complex than OSI layer 1, since it must provide for shared use, i.e. contention between the use of the D channel as a signalling channel, or as a data channel with the priority placed on its signalling capability. A contention layer is, therefore, incorporated in this layer functioning in a somewhat similar manner to CSMA/CD operation in a bus system.

An indication of the function of this control may be seen from the I430 frame structure shown in Fig. 9.4. This is 48 bits in length. The start of a frame is delineated by a deliberate violation of the alternative mark inversion coding, described in section 4.2.3. Apart from the bits used for transmission of the B, B' and D channel information, several other bits are included in the frame. One bit allows for contention resolution on the bus carrying the statistical multiplexing of a number of control signals on the D channel. The A bit shown is used as part of an activation sequence which puts the terminal into operation as required (by either network or terminal) and also to indicate the end of a successful activation sequence when both the network and the terminal are in synchronism. A D channel echo bit is used to indicate which terminal may proceed following a collision situation. A collision detector monitors the digital value in the D channel echo bit position. If a collision is detected by a change in the value of this bit, the terminal retransmits at a later random time. By controlling this 'back-off' period, a priority mechanism is provided that allows D channel signal messages to take a higher precedence than D channel packet data.

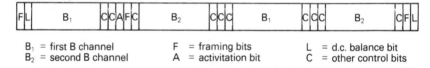

B₁ = first B channel   F = framing bits   L = d.c. balance bit
B₂ = second B channel   A = activitation bit   C = other control bits

Fig. 9.4 The I430 frame structure.

Layer 2 recommendations (I440 and I441) are based on the link access layer of X25 and provide similar functions, i.e.

a error detection using a 16 bit cyclic redundancy frame check sequence;

b error correction by retransmission of the defective frame;

c multiplexing of a number of call control procedures over the same link;

d flow control; and

e sequencing.

A further function is added to the ISDN layer 2 to provide for point-to-multipoint connection through allocation of layer 2 addresses to all relevant terminals connected to the interface.

Layer 3 recommendations (I450 and I451) specify the call control procedures used for the establishment of both circuit- and packet-switched calls, with the former being conveyed over the B channel and the latter using either the B or D channels. These procedures are quite different to X25 since a process of **out-band signalling** is followed. Since the incoming calls to the system will not contain sufficient information to identify a specific terminal connected to the passive NT2 bus, the call is conveyed to *all* terminals connected to the bus. Those terminals that are compatible with the calling terminal and free to answer will acknowledge the call, and the NT2 or the termination unit located in the exchange will carry out call allocation according to a pre-defined allocation algorithm. This makes the clearing procedure more complex than with in-band signalling but does enable the B channel to operate at higher efficiency since it will be free of the necessity to convey supplementary services such as tariff charge calculations.

The **primary rate access network** interface conforms to recommendation I421. This interface also refers to a subset of recommendations which are:

a   layer 1 recommendation, I431,

b   layers 2 and 3, as in recommendation I420.

The electrical characteristics of layer 1 are related to the PCM primary transmission rates of 2.048 or 1.544 Mbps specified in recommendation G703. Only point-to-point communication is applied, unlike the passive bus configuration in the basic rate access network interface.

This recommendation is especially important for interfacing to leased digital circuits. Three signals are transferred across the interface, a 64 kbps data signal, and two timing signals, one at 64 kHz for the clock frequency and the other at 8 kHz for octet timing. Three permissible implementations of timing control are available:

a   duplex timing in which the timing information is passed in both directions;

b   a centralized clock where a common external clock provides timing for both directions of transmission; and

c   simplex timing in which timing information for both directions of data transfer is passed one direction at a time.

For operation at 2.048 Mbps a coaxial pair is used for each direction of transmission using HDB3 coding. As discussed earlier (section 4.2.3), HDB3 is a method of encoding binary signals into a pseudo-ternary (three-state) signal arranged to ensure that no d.c. component is introduced into the transmitted signal. Either synchronous or asynchronous operation can be used.

9.1.7 Packet communication

CCITT recommendation I472 is concerned with handling of packet communication. Inter-networking between ISDN and PSS currently requires a two-stage calling procedure (Fig. 9.5). First, a call must be made across the ISDN exchange termination (ET), using 1472 procedure to an interworking port (IP) on the PSS network. Second, a call set up across PSS is required using PSS procedures to the destination address (Fig. 9.5.). Only access via the B channel of the ISDN is possible. An extended version of this connection is being developed which will provide a packet handling function with the ISDN. This will involve both channels B and D access and include processing for packet calls, path setting and X25 protocol compatibility.

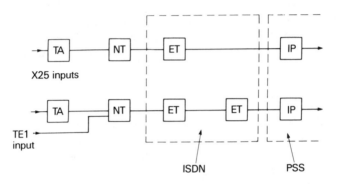

Fig. 9.5 PSS/ISDN operation.

9.1.8 Range of user services

Many different countries are currently initiating an ISDN service using the 144 kbps basic rate access service and plan to provide a number of services using 2B + D channels. These services will be considerably enhanced when the full primary rate access service is developed. A list of those proposed at present include:

Data and text services

telegraph	telex	teletex
telecommand	intercomputer data	viewdata
home newspaper	radio paging	alarms
electronic funds transfer		

Speech services

telephone	speechgram	speechdata
radiophone	hi-fi (music)	stereophonic
conference		

Still picture services

facsimile (line drawing, half-tone, colour)
picture viewdata   home newspaper   still hologram

Real time picture service

slow-scan TV   high definition TV   picturephone
confravision   colour TV   stereo TV
moving hologram

Since ISDN is a completely digital service from a user's premises there is a need for codecs or other analogue/digital devices to be incorporated directly at the transmitting end. The leading chip manufacturers, Motorola, A T & T, AMD and Intel have developed chips suitable for inclusion in the network terminal interface (NTI). Examples are the Intel 29C53 transceiver and the 29C48 programmable codec/filter. This latter is designed to convert analogue (speech) signals into digital form for transmission and vice versa, allowing the connection of a digital telephone to ISDN and thus simplifying the design of the NTI which presently forms an additional unit at the users' premises.

9.2 ELECTRONIC MESSAGING SYSTEMS

With the development of easily acceptable public data networks came the realization that they may be used not only to transmit data between different computer machines and systems but also directly from person to person in much the same way as we use the existing telephone and postal services. Each user has access to a particular area in a central computer store, known as his 'mailbox', where messages may be stored and accessed. Transfer of this information to another user's mailbox is a simple matter of data transfer to a known address in the computer store where the destination mailbox is located. Much activity in recent years has been directed towards devising suitable message systems and protocols compatible with existing national network arrangements for this data transfer and access. The areas of support looked for in defining a message system include:

a support for any type of message, e.g. text data, speech, graphics, facsimile etc;

b message interchange across public and private message systems;

c connection to teletext, facsimile and videotext systems;

d support for personal computers and work stations connected to private and public data networks; and

e support for private electronic messaging systems using distributed message exchanges.

In order to be compatible with existing systems and those under development it is also necessary that the messaging system be compatible with the seven-layer ISO protocol and the relevant data passing protocols.

9.2.1 The CCITT X400 recommendation

An agreed specification that meets these requirements is the CCITT X400 series of recommendations, defining a number of services with formats and protocols to support them. These facilities are called a **message handling system** (MHS) and include a **message transfer system** (MTS) and an **interpersonal message system** (IPMS). The system model for MHS conforms to the ISO application layer 7 protocol and consists of a number of sublayers as shown in Fig. 9.6.

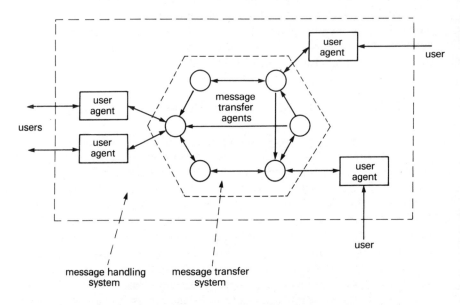

Fig. 9.6 System model for a message handling system.

The **User Agent** (UA) sublayer accepts information from the user concerning such matters as message identification, message subject, address of primary/secondary recipients, data and level of priority. Several separate UAs can be employed. The UA provides a common interface between a defined X400 system and the undefined local user environment which could form part of an existing mailbox service, e.g. Telecom Gold, One-to-One, Comet, Telemail and other systems defined by a content protocol, X420. This protocol is also the agent which provides interconnection between the different mailbox services so that a user wishing, for example, to transfer a message out of his Telecom Gold mailbox to one located in a One-to-One mailbox may, in principle, be able to do so.

The second sublayer, the **message transfer agent** (MTA), transmits and receives messages on behalf of one or more UAs. It analyses the address to determine the destination for the message, acts as a sorting office and arranges recorded delivery or other services requested by the message. (One important service would be the ability to transmit a telex message using the MTA.) Finally, it prepares the message for transmission to the next lower level of the ISO system, the presentation layer, and hence to a packet-switched carrier such as PSS or ISDN. It does this through a **reliable transfer service** (RTS) complying with CCITT recommendation X410[6].

In using a messaging system, a number of separate activities are associated with the user's mailbox, such as message composition, preparing the 'envelope' or address of the message, facilities for the accessing and manipulation of the mailbox contents and finally transmitting the message and obtaining an acknowledgment. These tasks of the MHS are shared between the central computer system maintaining the user's mailbox location, and the user's terminal where messages are prepared for transmission and recovered for inspection. The terminal may comprise simply a connection to a time-shared computer, having access to a text editor located in the distant computer, or it may be a personal computer containing its own word-processing software and a screen-based editor. Most message systems work with unstructured text, but the envelope format is rigidly fixed by the user agent protocol. Envelope composition is usually carried through some form of interactive prompting by the system to ensure that the format is complied with. User commands, apart from those used in message composition and addressing (envelope), include those for locating specific, messages, e.g. GO TO (message), NEXT, PREVIOUS or string searches such as MESSAGES FROM (number) TO (number), or between dates, e.g. DATE BEFORE (date), DATES BETWEEN $(date_1, date_2)$ etc., together with manipulative commands such as MOVE (number), PRINT (list), DELETE, FORWARD (address) etc. Multi-media editors are also being developed which will deal with the transmission of graphics and other data and these too will add to the user commands available[7].

Of particular importance are the conventions used for naming both the physical sites and mailboxes. Earlier systems, such as Arpanet mail used a simple two-level addressing structure with a single address for the storage computer (the host) and the mailbox. Now, several different forms of hierarchical and distributed naming methods are in use, and interaction between various individual mailbox systems is needed. As a consequence it is often necessary to carry out complex name manipulation between different message systems to convey information across them. The position is slowly changing, however, and as more mailbox systems accept the need for common standards such as X400, then the user will realize a greater 'transparency' in the mail services offered.

The CCITT recommendations are, at present, less detailed than other older communication protocols that have now been accepted as standards by the industry. It does not, for example, define a complete MHS implementation. Instead, a system model is defined together with a set of elements to be supported, and standard message formats and protocols to support them. A *de facto* inter-message service has presently been defined by the UK Department of Trade and Industry[8]. This standard defines the format of messages passing between the user and the UA with the encoding method defined in CCITT recommendation X409. This includes preceding each field with one or more bytes of information defining the type and length of the following field. This allows the MTA to interpret the onward transmission for the message including the possibility of inter-message system transmission —at least in the UK.

9.3 MANUFACTURING APPLICATIONS PROTOCOL

The past decade has seen a large increase in the use of computer-controlled machines for the manufacturing and process control industries. This is due to the availability of small, low-cost computers having considerable processing power compared with similar equipment a few years ago. A single manufacturing organization may have a number of these operating at shop floor level and elsewhere, each designed for a specific task and generally incompatible with each other. There remains a need, however, to integrate their operations in the interests of organizational and manufacturing efficiency. The automobile industry was one of the first to adopt computer control of production processes and was the first to experience the problems associated with lack of easy communication between the different machines used (quite often supplied from different vendors). As a natural consequence the concept of a single communications protocol covering the special needs of automation was first developed by the automobile industry.

The technique is referred to as **manufacturing applications protocol** (MAP) and is the result of several years of activity by a working group initated by General Motors in the early 1980s. MAP is a set of communications protocols applicable to a wide range of multi-vendor equipment and based on the ISO seven-layer system. Whilst operating as a local area network, its software supports all seven layers of the ISO model and is therefore considerably more flexible in operation than the LAN systems discussed earlier in Chapters 7 and 8.

9.3.1 Choice of LAN operation

The specific requirements for a LAN to operate at manufacturing shop floor level are quite stringent and include:

1 Support of a number of different types of devices.

2 Provision for a range of services on the same network, e.g. speech, data, vision and control.

3 Real-time communication.

4 Communication between complex computer devices rather than between computer and terminal/peripheral.

5 High reliability and absence of error to minimize faulty manufacture of machined or assembled units.

6 Operation in a harsh environment, i.e. dirty, high level of electrical interference, supply voltage fluctuations etc.

These requirements effectively preclude the use of bus protocols operating under CSMA/CD. The risk of repeated data collision is too great, and futhermore its efficiency deteriorates under high traffic loading conditions. The choice lies between a ring or token bus system protocol, either of which can operate under more controlled, i.e. deterministic, conditions.

A broadband token bus system arranged as a logical ring has been chosen by the MAP team for a number of reasons governed by a need to meet the flexibility demanded by the industry. Some of these are:

a the need to allow several devices to communicate at the same time;

b a requirement for priority message operation;

c no minimum packet length;

d reliable operation with high traffic loading; and

e calculable maximum waiting time for a given station.

An additional advantage of a broadband system in this context is that it becomes possible to set up a test network to run in parallel with a live one, using different channels in the same environment. This enables system development work to be carried out on a MAP network without interfering with live running on another channel.

Connection between stations on the LAN is made through baseband coaxial cable or fibre optic cable. The latter, although not yet widely used for this purpose, is attractive since it has the properties of high bandwidth, immunity to electrical interference, vibration, and a corrosive chemical atmosphere, all of which may be present in an industrial environment.

9.3.2 MAP Standards

Whilst the ISO and standards committees of the IEEE are developing together widely acceptable recommendations for manufacturing shop floor communication, the group of protocols known as MAP is the responsibility of the General Motors working group. Interaction between this group and the standards committees means that the final recommendations, when issued by ISO and the IEEE, are likely to be very close to the *de facto* standards already in use by a number of manufacturing concerns including General Motors, Ford Motor Company, Chrysler, McDonald Douglas and many others. MAP-compatible products to these standards are also available from some of the larger electronic concerns such as Digital Equipment, Hewlett-Packard, Intel and Motorola.

MAP, as at present designed, is a broadband token bus system using community television coaxial cable as the physical medium. The data rate is at 10 Mbps, which is capable of being shared as a number of small sub-frequency bands to provide separate data channels.

9.3.3 MAP protocols

The standards adopted for MAP closely follow the ISO model as can be seen from their comparison in Fig. 9.7. The lowest (physical) layer 1 conforms to the IEEE 802.4 token bus scheme discussed in section 8.5.3. Of the three

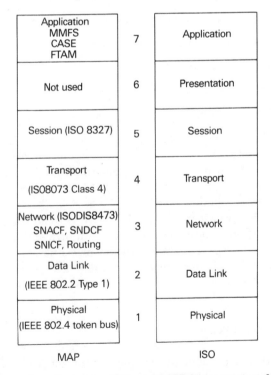

Fig. 9.7 Comparison between ISO and MAP layer protocols.

modulation techniques offered by this layer, MAP applies QAM operation at a rate of 10 Mbps. Two-channel operation is applied with data transmission and data reception carried on two different frequency bands. Transmission from the network elements is generally on the lower frequency with the higher frequency used for reception. The signals are carried through the coaxial media to a headend which contains a remodulator for frequency conversion. This accepts the lower frequency and converts it to a higher frequency transmission as required at the receiving end.

The data link layer 2 applies the IEEE 802.2 standard. This defines two alternatives: connectionless (Type 1) and connection-oriented (Type 2) operation. MAP uses Type 1 which is able to detect and discard messages having incorrect parity, leaving to the higher layers the detection of missing or duplicated messages. A third type of layer 2 operation has been proposed by the IEEE which includes an immediate acknowledgement, and this could be used where small subnets are applied within MAP rather than applying them to the entire network. The MAC method chosen for this layer is token-passing using a bus configuration.

The network layer 3 applies the connectionless network service which detects missing message fragments via a time-out mechanism (It will be remembered from our discussion of ring networks in Chapter 7 that this is a general problem with repeater networks forming a real or logical loop.) This layer is important to MAP since the ability to access devices through intermediate networks is fundamental to the ideas of MAP. Due to the size of manufacturing installations it is seldom practical to place all the systems in one common link. Conceptually, the ISO divides this layer into four sublayers:

1 subnetwork access facility (SNACF);

2 subnetwork-dependent convergence facility (SNDCF);

3 subnetwork-independent convergence facility (SNICF); and

4 Routing and relaying.

These sublayers are concerned with converting global address information into routing information, maintaining message routing tables and algorithms and switching each incoming message to its correct outgoing path.

The transport layer 4 carries out detection of complete missing or duplicated messages and provides a network-independent transport service to the session layer 5. This layer enhances the transport service with methods of managing and structuring a dialogue between two systems using the data transport provided on a transport connection. The session layer 5 provides the services required by the application layer and is at present limited to the ISO session ISO8327 protocol which only supports a subset of the standard but is considered adequate for the application layers supported. Whilst the ISO standard allows for simplex, half-duplex and full duplex connection only full duplex connection is currently supported by MAP layer 5. The current version of MAP 2.1 does not support a presentation layer 6, so that the session and application layers of the model form the highest layers of the MAP protocol.

The applications layer 7 supports three main service divisions: common applications service elements (CASE), file transfer applications and management (FTAM) and a manufacturing message format standards (MMFS). CASE is intended to provide capabilities which are required by application processes for open systems interconnection independent of the application. Examples include gateway transmission and the connection of programmable devices. FTAM is specified for both MAP and TOP (and is currently the only protocol supported at level 7 of TOP). This deals with procedures for moving files from one open system to another together with file access and file security. MMFS is intended as a common versatile mechanism for machine-independent process-to-process information exchange. It may be applied, for example, to managing the interconnections between small programmable controllers and robots which are not covered by the ISO FTAM file transfer scheme.

Various MAP-compatible end-user products are now being developed. These include MAP interface boards which can accommodate a number of different vendor mainframes, MAP bridges which can connect two MAP bus networks together for the purpose of extension or isolation and which operate at the data link level, MAP routers to join together two dissimilar networks via the ISO network protocol, and MAP gateways to link MAP networks to non-OSI subnetworks. These gateways allow access at the applications level only, so that the subnetwork plays no part in the MAP distribution control system but is simply capable of transferring data between the two networks. To support terminals and personal computers at shop floor level, MAP services are also available to provide high-speed MAP connection and RS-232c terminal connection.

Alternative MAP specifications are also becoming available which deal with two special conditions of equipment: interconnection and control.

1 **Carrierband MAP**: A one-channel system operated at 5 Mbps which is simpler and cheaper to implement at layer 1, with other layers remaining unchanged. This might be used, for example, to connect a sensor in a process control cell to the system.

2 **MiniMAP**: A layered architecture consisting of layers 1, 2, and 7 of the full MAP protocol. This allows rapid data transfer but is not OSI-compatible since it is incapable of accessing devices operating the full seven-layer architecture.

Uses for these three versions have been suggested as:

a Full MAP to provide factory-wide communication in the manufacturing and process control industries.

b   Carrierband MAP to provide a cell-level communications system in the manufacturing industry.

c   MiniMAP for time-critical operation in the process control industry.

Development of MAP standards and protocols is an on-going activity. Version 3.0 of MAP (when issued) will differ from MAP 2.1 described above in a number of important respects, including the file transfer arrangements, real-time messaging and inclusion of a presentation layer (which is missing in the current standard). The file transfer protocol, FTAM, will be enhanced to permit selective retrieval and update of one record at a time (instead of bulk file transfer as at present). The new presentation layer will provide negotiation for the representation of data for use by the applications layer 7. Here translation between different encoding systems (DEC, IBM, Hewlett-Packard etc.) would be undertaken. Additionally, it will allow FTAM to work for all types of files and thus enhance its management capabilities. The MMFS system included in layer 7 will be replaced by a new standard, known as RS-511 which will allow implementation to the CCITT X400 message recommendations and thus introduce a useful interface to the messaging systems in general use and described earlier in this chapter. It is likely that these improvements, together with the ability of the semiconductor manufacturers to supply single-chip LAN controllers to MAP specifications, will also promote the general development of LAN communication protocols, particularly in inter-networking with other more general types of network[9].

9.4   TECHNICAL AND OFFICE PROTOCOL

Whilst MAP protocols need to be deterministic—one shop floor device must be able to communicate with another within a known period of time — office routines can accomodate small and variable delays in data transmission. For this reason, the **technical and office protocol** (TOP) has standardized on the cheaper CSMA/CD protocol using the IEEE 802.3 standard at level 1 which already forms the basis of a large number of office networks that have installed Ethernet LANs. The two systems, MAP and TOP, however, employ substantially the same standards at the higher layers and are, therefore, essentially able to interconnect with each other and so achieve better communication between administration, design and production than may be obtained with separate network systems.

TOP is a development parallel to MAP, designed for the office, and was originally initiated by Boeing Computer Services in the USA. In a practical application, designers and technicians carry out development work using systems on the TOP network, passing their results across to the system to the MAP network for implementation and production. To do this, a device known as a **router** (also known as an intermediate open system) is used to connect multiple networks together at a common point (Fig. 9.8). A router uses the lower three OSI layers to provide path selection and alternate routing. Unlike a bridge connecting two dissimilar networks, this connection is not a transparent one and it is necessary to provide destination network layer addresses and status of the connected networks. The routing service provided by the router includes finding paths between networks having different address domains and frame sizes, a situation which exists when attempts are made to interconnect the MAP and TOP networks.

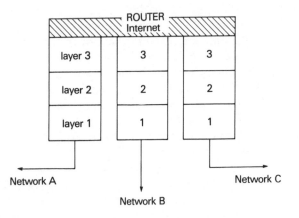

Fig. 9.8 A router protocol connection.

TOP version 1.1 uses 50 ohm Ethernet coaxial cable on an omnidirectional 10 Mbps bus. The access technique imposes a limitation on the physical size of the network as a result of the round-trip propagation delay for a collision detection procedure. The maximum length of a cable segment is 500 m. This is generally adequate for an office environment, but is shorter than that possible with the MAP broadband system.

Whilst ISO protocol layers 3–6 are identical for MAP and TOP, layers 1, 2 and 7 are different. As mentioned previously, TOP uses IEEE 802.3 for layer 1 and the MAC sublayer 2, the data link layer. The services offered at layer 7, the application layer, reflect the office/design environment and include graphics, electronic mail and word/text processing. The relevant file transfer protocol for this layer using TOP is ISO FTAM8571.

As with MAP, the TOP system is also under development and when TOP 3.0 is released it will contain full X400 messaging protocol to layer 7 and an ability to exchange graphical information in accordance with international standards, including, possibly, 3-D graphics representation. Consideration is also being given to supporting the thinner 'cheapernet' coaxial cable for the Ethernet system which was mentioned in Chapter 7, together with the inclusion of a MAC layer supporting the IEEE 802.5 token ring network.

**SUMMARY
Chapter 9**

Electronic communication can take several forms (e.g. speech, data, vision, facsimile etc.), each of which generally requires a separate communication channel. There are advantages in developing a combined service which can accommodate all of these for transmission over one channel. One such service which has reached international agreement is the CCITT Integrated Services Digital Network (ISDN). This supports transmission over a single channel of two 64 kbps data channels and a 16 kbps common signalling channel and can accommodate several forms of digitized information from a composite digital signal transmitted directly from the users' premises. An extension of this technique can provide transmission of a wider bandwidth to permit multiple channels, vision and other wideband signals to be accepted.

Other convergence techniques that have been developed in recent years are electronic messaging systems (EMS) to convey written data directly between personal users, manufacturing applications protocol systems (MAP) and technical and office protocol (TOP).

These latter systems provide a technique for linking together a number of computer-controlled machines for the manufacturing and process control industries and for coordinating these with the associated design and administrative procedures. This enables the complete process of manufacture to be coordinated, with consequent improvement in efficiency and reduction of faulty manufactured and assembled units.

All of these convergent techniques use a common protocol model based on the ISO seven-layer model which makes the task of accommodating different information requirements and different computer vendors within a single system much easier to accomplish.

**PROBLEMS
Chapter 9**

P9.1   a In the I430 frame structure its commencement is indicated by deliberate violation of the alternate mark inversion coding. How can this procedure enable the start of the frame to be identified?

b How may a contention bit in the D channel frame be used to control the allocation of this channel to the appropriate B or B' channel?

P9.2 Give some reasons for the adoption of a seven-layer protocol layer model in the development of MAP. Why is the operation of a token bus protocol preferred against a contention protocol for link layer communication?

P9.3 Distinguish between the message transfer agent and the user agent in the operation of an X400 electronic message system.

P9.4 A private organization has a CBX unit capable of operating from the primary rate access system. The equipment served by this unit includes:

80 telephone channels each of 64 kbps,
3 slow-scan televevision channels with the picture repeating every 2 seconds,
50 data channels of 2.4 kbps of high priority, and
150 data channels of 1.2 kbps of low priority.

What is the minimum number of primary rate access channels required and how should they be allocated to the different requirements?
(A 600-line television picture can be assumed with an aspect ratio of 4 : 5).

P9.5 Give reasons for the choice of different LAN protocols for MAP and TOP. Why is the same system not chosen for the two protocols since they need to work together?

10

RECOMMENDED READING

10.1 COMMUNICATIONS LITERATURE

The books on data and computer communications can vary considerably in the extent and range of subject matter and not all of these are suitable for undergraduate study. They also tend to 'date' quite rapidly as new means are developed, supplanting earlier less adequate techniques. Good general introductory texts to the subject at the basic communication level are:

Martin, J. (1981). *Telecommunications and the Computer*. Prentice-Hall, Engelwood Cliffs.
National Computing Centre (1975). *Handbook of Data Communications*. NCC Publications, Manchester.
O'Reilly, J.J. (1984). *Telecommunication Principles*. Van Nostrand Reinhold, UK.
Roden, M.S. (1982). *Digital and Data Communication Systems*. Prentice-Hall, Englewood Cliffs.
Stallings, W.S. (1985). *Data and Computer Communications*. Macmillan, New York.

These all include a good component of the physical and electrical engineering principles involved in the foundations of the subject.

More theoretical texts concerned with network design and operation are:

Davies, D.W. and Barber, D.L.A. (1973). *Communication Networks for Computers*. John Wiley, London.
Halsall, F. (1985). *Data Communications and Computer Networks*. Addison-Wesley, Wokingham, UK.
Tanenbaum, A.S. (1986). *Computer Networks*. Prentice-Hall, Englewoood Cliffs.

Others are concerned almost entirely with communications protocols and flow control and are thus able to deal very adequately with a complex network situation. Examples are:

Davies, D.W., Barber, D.L.A., Price, W.L. and Solomonides, C.M. (1980). *Computer Networks and their Protocols.* John Wiley, Chichester.

Seidler, J. (1983). *Principles of Computer Communication Network Design.* John Wiley, New York.

Stallings, W.S. (Editor) (1985). *Tutorial: Computer Communications, Architecture, Protocols, and Standards.* IEEE Press, New York.

A number of specialized publications are of interest to the communications manager who needs to be aware of the range of techniques available without too much technical detail but with the emphasis on comparative performance between techniques. Examples are:

Bennett, G.H. (1985). *Pulse Code Modulation.* Marconi Publ., Chelmsford, UK.

Black, U.D. (1983). *Data Communication Networks and Distributed Processing.* Reston Publ., Virginia, USA.

Flint, D.C. (1983). *The Data Ring Main.* John Wiley, Chichester.

Gee, K.C.E. (1983). *Introduction to Local Area Computer Networks.* Macmillan, Basingstoke, UK.

Stuck, B.N. and Arthur, E. (1985). *Computers and Communication Network Performance Analysis Primer.* Prentice Hall, Englewood Cliffs.

Wushow, C. (Editor) (1983). *Computer Communications Vol 1 — Principles.* Prentice-Hall, Englewood Cliffs.

Finally, there are a number of detailed texts used for reference or study on particular aspects of the subject, such as fibre optics or satellite communication. Some recommended texts are:

Dallas, I.N. and Spratt, E.B. (Editors) (1983). *Ring Technology Local Area Networks.* North Holland.

Feher, K. (1981). *Digital Communications — Satellite/Earth Station Engineering.* Prentice-Hall, Englewood Cliffs.

Gowar, J. (1984). *Optical Communications Systems.* Prentice-Hall, Englewood Cliffs.

Siemens Ltd (1983). *Optical Communications — A Telecommunications Review.* John Wiley, Chichester.

A useful source of up-to-date information on digital communications subjects are contained in the reports and publications of the **National Computing Centre**, Manchester, UK. New titles are continually being added to their series which includes most of the subjects considered in this book.

10.2 JOURNALS

IEEE Transactions on Communications
IEEE Communications Magazine (IEEE Computer Society)
IEE Communications, Radar, and Signal Processing
British Telecom Technology Journal
British Telecom Engineering
Computer Communications, Butterworth
Computer Standards and Interfaces, North Holland
Computer Networks and ISDN Systems, Elsevier

10.3 CONFERENCES

A wide range of conferences and symposia covering the subject of computer communications are presented each year by the learned societies, the IEEE, IEE, IERE etc. as well as by a number of commercial organizations such as INFOTECH and ONLINE. Details of these are to be found in the technical press.

Overseas conferences are numerous. An important event in telecommunications is the series of World Telecommunications Conferences which take place every few years under the auspices of the International Telecommunications Union in Geneva.

10.4 STANDARDS COMMITTEES

These issue definitive standards and reports which are referred to by the communications industry. The most important of these is the **Comité Consultatif Internationalé de Télégraphie et de Téléphonie** whose address is: CCITT, Place des Nations, CH-1211 Geneva 20, Switzerland. (The same address applies for the International Standards Organization (ISO).)

CCITT publishes formal standards recommendations at approximately yearly intervals. Some recent recommendations related to the topics raised in this book are a set of volumes known collectively as the 'Red Book' which contain the recommendations of the 8th Plenary Assembly which took place during 1984 in Geneva. Recommendations relevent to computer communications are contained in Volume VIII and consist of the following seven sections:

VIII.1   Data communication over the telephone network.
           Series V recommendations.

VIII.2	Data communication networks: services and facilities. Recommendations X1 – X15.
VIII.3	Data communication networks: interfaces. Recommendations X20 – X32.
VIII.4	Data communication networks: transmission, signalling and switching, network aspects, maintenance and administrative arrangements. Recommendations X40 – X181.
VIII.5	Data communication networks: OSI, system description techniques. Recommendations X200 – X250.
VIII.6	Data communication networks: inter-networking between networks, mobile data transmission systems. Recommendations X300 – X353.
VIII.7	Data communication networks: message handling systems. Recommendations X400 – X430.

They can be ordered from the CCITT Sales Service at the address given above.

Another important body is the **IEEE Standards Board** whose address is: IEEE Standards Board, 345 East 47th Street, New York 10017, USA.

Publications describing their IEEE 802 standards are available from Wiley-Interscience Publishers. Current titles are:

IEEE 802.1 Relationships between standards.

IEEE 802.2 Logical Link Control Protocol.

IEEE 802.3 CSMA/CD.

IEEE 802.4 Token-passing Bus.

IEEE 802.5 Token-passing Ring.

Other bodies from which standards documents may be obtained are:

British Standards Institute (BSI), Sales Department, Linford Wood, Milton Keyes MK14 6LE, UK.

American National Standards Institute (ANSI), 1430 Broadway, New York, NY 10018, USA.

National Bureau of Standards, Technical Information and Publications

Division, Washington DC 20234, USA.

Information Technology Standards Unit (ITSU), Department of Trade and Industry, 29 Bressenden Place, London SW1E 5DT, UK.

The Institution of Electrical Engineers in London publish yearly a comprehensive survey of networking standards with addresses and details of the current networking committees, compositions and status of definitions agreed. This is entitled: *Worldwide Standardisation Activities on Open Systems Interconnection and Local Area Networks* and is obtainable from: The Institution of Electrical Engineers, Savoy Place, London WC2R 0BL, UK.

10.5 CHAPTER REFERENCES

10.5.1 Chapter 1

1. Young, P. (1983). *Power of Speech*. George Allen & Unwin, London.
2. Atkinson, J. (1950). *Telephony*. Pitman, London.
3. Flowers, T.H. (1979). Penetration of electronics into telephone switching. *Proc.IEE* **106b**, Suppl.15, 901.
4. Vanner, N.J. (1979). Architecture of System X — the digital trunk exchange. *Post office Electrical Engineering Journal*, **72**, 142–148.
5. CCITT 'White Book' (1983). Vol. VIII.15. ITU, Geneva.
6. Hughes, C.J. (1986). Switching — state of the art. *Br. Telecom Technol. J.* **4**, (1), 5–19.
7. Breary, D. (1974). A long term study of the United Kingdom trunk network. *POEEJ* **66**, 210—216.
8. Murray, W.J. (1982). The emerging digital transmission network. *Br. Telecom Eng J.* **1**, 166.
9. Jones, W.G.T., Kirtland, J.P. and Moore, BL.W. (1979). Principles of System X. *POEEJ* **72**, 75–80.

10.5.2 Chapter 2

1. Bennet, G.H. (1983). *Pulse Code Modulation and Digital Transmission*. Marconi Instruments Publ., Chelmsford, UK.
2. CCITT 'Orange Book' (1977). *Data Transmission over the Telephone Network*. Vol. VIII.1, pp.79–84. I T U, Geneva.
3. Freeman, R. (1981). *Telecommunications Transmission Handbook*.

John Wiley, New York.

4. Bell Telephone Laboratories (1982). *Transmission Systems for Communications*. Bell Labs., USA.

5. Lucky, R.W., Salz, J. and Weldon, J.R. (1965). *Principles of Data Communication*. McGraw-Hill, New York.

6. Lathi, B.P. (1968). *Communication Systems*. John Wiley, New York.

7. Shannon, G.E. (1948). A mathematical theory of communication. *Bell System Techn. J.* **27**, 623–656.

8. Oetting, J. (1979). A comparison of modulation techniques for digital radio. *IEEE Trans. on Communs* **COM-27**, (12), 1752–1762.

9. CCITT 'Yellow Book' (1980). *Data Communication over the Telephone Network*, Vol VIII.1. ITU, Geneva.

10. Rose, K.R. and Forse, N.J.A. (1985). Speech echo control in the UK Network. *IEE 3rd Conf. 'Telecom Trans.'*, pp.196–199, March.

11. Chu, W. (1973). Asynchronous time-division multiplexing systems. In *Computer Communication Networks*, Ed. Abramson, N. and Kuo, F. John Wiley, New York.

10.5.3 Chapter 3

1. UK Post Office (1975). *Handbook of Data Communications*. NCC, Manchester, UK.

2. Bennett, G.H. (1983). *Pulse Code Modulation and Digital Transmission*. Marconi Instruments Publ., UK.

3. Hooper, R.C. (1985/6). The development of single-mode fibre transmission system at BTRL, Parts I and II. *Br. Telecom Eng.* **4**, 74 and 193–198.

4. Adams, C.J. (1984). The universe project. In *Information Technology and the Computer Network*, Ed. Beauchamp, K.G. Springer-Verlag, Heidelberg.

5. Manassah, J.T. (1982). *Innovations in Telecommunications Part.B.* Academic Press, New York.

6. Lathi, B.P. (1968). *Communication Systems*. John Wiley, New York.

7. Davenport, W.B. and Root, W.L. (1958). *An Introduction to the Theory of Random Signals and Noise*. McGraw-Hill, New York.

8. Byrne, C.J., Karafin, B.J. and Robinson, D. (1963). Systematic jitter in a chain of digital regenerators. *Bell System Techn. J.* **42**, 2679–2714.

9. Zegers, L.E. (1967). The reduction of systematic jitter in a transmission channel with digital regenerators. *IEEE Trans. Commun. Technol.* **COM-15**, (4), 542–551.

10. Nyquist, H. (1928). Certain topics in telegraph transmission theory. *Trans. AIEE* **47**, 617–644.

11. Lucky, R.W., Salz, J. and Weldon, E.J. (1968). *Principles of Data Communcation*. McGraw-Hill, New York.

12. Lucky, R.W. and Rudin, H.T. (1966). Generalized automatic equalization for communication channels. *Digest Techn. Papers IEEE Int. Communs. Conf.* 22.

10.5.4 Chapter 4

1. Widrow, B. (1961). Statistical analysis of amplitude quantised sampled data systems. *IEEE Trans. Appl. Indust.* **ID-52**, 555–570.
2. Bell Telephone Laboratories (1982). *Transmission Systems.* Murray Hill, New Jersey.
3. CCITT 'Green Book' (1973). *Line Transmission*, Vol. III. ITU, Geneva.
4. Croisier, A. (1970). Introduction to pseudo-ternary codes. *IBM J.* **14**. (4), 354.
5. Martin, J. (1981). *Computer Networks and Distributed Processes.* Prentice-Hall, Englewood Cliffs.
6. Bennett, G.H. (1983). *Pulse Code Modulation and Digital Transmission.* Marconi Instruments Publ., UK.
7. Bhargava, V. (1983). Forward error correction schemes for digital communication. *IEEE Communs*, Jan.
8. Martin, J. (1970). *Teleprocessing Network Organisation.* Prentice-Hall, Englewood Cliffs.
9. CCITT 'Yellow Book' (1981). *Data Communications over the telephone network*, Vol. VIII.8.1. ITU, Geneva.

10.5.5 Chapter 5

1. Martin, J. (1976). *Telecommunications and the Computer.* Prentice-Hall, Englewood Cliffs.
2. Kleinrock, L. (1976). *Queueing Systems; Vol. 2, Computer Applications.* Wiley-Interscience, New York.
3. Price, W.L. (1977). *Data Network Simulation: experiments at the National Physical Laboratory, 1968–76.* Computer Networks, 1, 199.
4. Reinde, J. (1977). Routing and control in a centrally directed network. *AFIPS Conf. Proc. Natn. Comput. Conf.* 46, 608–630.
5. Fultz, G.L. (1972). *Adaptive routing techniques for message switching computer communication networks.* University of California, Los Angeles, Report UCLA-ENG-7352.
6. Bartlett, K.A., Scantlebury, R. A. and Wilkinson, P.T. (1969). A note on reliable full duplex transmission over half-duplex links. *Commun. ACM* **12**, 260.
7. Davies, D.W., Barber, D.L.A. *et al.* (1984). *Computer Networks and their Protocols.* John Wiley, Chichester, UK.
8. Stallings, W. (1985). *Data and Computer Communications.* Macmillan, New York.

9. Beauchamp, K.G. and Yuen, C.K. (1979). *Digital Methods for Signal Analysis*, Chapter 2. George Allen & Unwin, London.
10. Kleinrock, L. (1976). *Queueing Systems Vol II — Computer Applications*. Wiley, New York.

10.5.6 Chapter 6

1. Davies, D.W., Barber, D.L.A., Price, W.L. and Solomonides, C.M. (1979). *Computer Networks and their Protocols*. John Wiley, Chichester.
2. Tippler, J. (1979). Architecture of System X. *POEEJ* **72**, 138–141.
3. Harris, L.R.F. and Davies, E (1979). System X and the evolving UK telecommunications network. *POEEJ* **72**, 2–8.
4. Roberts, L.G. (1970). Computer network development to achieve resource sharing. *AFIPS Conf. Proc. Spring J. Comput. Conf.* 36, 543.
5. Heart, F. (1970). The interface message processor for the ARPA computer network, *Ibid*, 551.
6. Kahn, R.E. and Crowther, W.R. (1972). Flow control in a resource sharing computer. *IEEE Trans. on Communs* **COM-20**, 539–546.
7. IEE Colloquium (1985). *The JANET Project*, Savoy Place, London WC2R 0BC, 26th November.
8. Smith, I.L. (1985). Operation of the JANET network. In ref. 7 above.
9. Report (1985). *Future facilities for advanced research computing*. SERC, Swindon, ISBN 0 901 660 73 6.

10.5.7 Chapter 7

1. Dubery, J.M. and Bartram, D (1986). The CLEARWAY ring as a tool in a university environment. *IEE Colloquium*, 'Computer Networks in Education', Savoy Place, London, 20th January.
2. Digital Equipment Corp., Intel Corp. and Xerox Corp. (1980). *The Ethernet: A Local Area Network Data Link Layer and Physical Layer Specification*. Manufacturers' report.
3. Bender, R., Abramson, N., Kuo, F.F., Okinaka, A. and Wax, D. (1975). Aloha packet broadcast — a retrospect. *Proc. Natn. Comput. Conf.*, 203.
4. Stallings, W. (1984). *Local Networks: an Introduction*. Macmillan, London.
5. Pierce, J.R. (1972). Network for block switches of data. *Bell System Techn. J.* **51**, 1133–1175.
6. Wilkes, M.V. and Wheeler, D.J. (1979). *The Cambridge Digital Communication Ring*. Proc. Local Area Communication Network Symposium, Boston, May.

7. Rubinstein, M.J., Kennington, C.T. and Knight, G.J. (1981). Terminal support in the Cambridge ring. *ONLINE Conference*, London, pp. 475–490.
8. Temple, S. (1984). The design of the Cambridge fast ring. In *Ring Technology — Local Area Networks*, Eds Dallas, F. and Spratt, E.B. North Holland.
9. Dixon, R.C. (1982). Ring network topology for local data Communications. *IEEE COMPCON Fall Proc.*, Washington, pp. 591–605.
10. Bux, W. (1983). *Local Area Subnetworks — a Performance Comparison*. IBM Zurich Research laboratory, Switzerland.
11. Penny, B.A. and Baghdadi, A.A. (1979). Survey of communication loop networks. *Comput. Communs*, **2**, Aug./Oct.
12. Stevens, R.W. (1983). MACROLAN: A high performance network. *ICL Techn. J.* **3**, 289–296.
13. Bates, R.J.S. and Saner, L.A. (1985). Jitter accommodation in token passing LANs. *IBM J. Res. Devel.* **29**, (6), 580–586.
14. Hafner, E.R., Nenadal, Z. and Tscharz, M. (1974). A digital loop communication system. *IEEE Trans. Communs.* **COM-22**, (6), 887–881.
15. Carlo, J.T. and Samsen G.R. (1986). Chip set points the way to token ring access. *Min/Micro Systems* **19**, (7), 121–127.
16. Smythe, C. and Spracklin, C.T. (1983). A high speed local area networking using spread-spectrum code division multiple access technique. *FOC/LAN 83 Proc.*, Atlantic City, New Jersey, pp.131–134.
17. Beauchamp, K.G. and Yuen, C.K. (1979). *Digital Methods for Signal Analysis*. George Allen & Unwin, London.

10.5.8 Chapter 8

1. Davies, D.W., Barber, D.L.A., Price, W.L. and Solominides, C.M. (1983). *Computer Networks and their Protocols*. John Wiley, Chichester.
2. McCrum, W.A. (1984). Open system interconnection. In *Information Technology and the Computer Network*, Ed. Beauchamp, K.G. Springer-Verlag, Berlin.
3. Carlson, D.E. (1980). Bit-oriented data link control procedures. *IEEE Trans. on Communs*, **COM-28**, (4), 455–467.
4. CCITT X series recommendations (1984). *The Red Book*, Vols VIII.2 and VIII.3. ITU, Geneva.
5. Marsden, B.W. (1986). *Communication Network Protocols*. Chartwell Bratt, UK.
6. *A Network-independent Transport Service (Yellow Book)* (1980). Prepared by Study Group 3 of the PSS users' forum, SG3/CP(80).
7. *A Network-independent File Transfer Protocol (Blue Book)* (1981).

Prepared by the FTP Implementors' Group of the Data Commun. Protocol Unit, NPL, Teddington, Middlesex FP-B(80).

8. *A Network-independent Job Transfer and Manipulation Protocol (Red Book)* (1981). Prepared by the JTMP working party of the Data Commun. Protocol Unit, DCPU/JTMP (80), (address as in ref. 7).

9. Bennett, C.J. (1982). *The JNT Mail Protocol (Grey Book).* Dept of Computer Science, University College, London.

10. Metcalf, R.M. and Biggs, D.R. (1976). 'Ethernet': Distributed Packet switching for local computer networks. *Communs of the ACM* **19**, (7), 395–403.

11. Kleinrock, L. and Tobagi, F. (1975). Random access techniques for data transmission over packet-switched radio channels. *Proc. Natn. Comput. Conf.*, pp.187–201.

12. Sharpe, W.P. and Cash, A.R. (Eds) (1982). *Cambridge Ring Interface Specifications (Orange Book)* prepared by SERC and the JNT (address as in ref. 7).

13. Clancy, G.J., *et al.* (1982). The IEEE 802 Committee States its Case Concerning Its Local Network Standards Efforts. *Data Commun.*, April.

14. Sze, D.T.W. (1984). IEEE LAN Project 802 — A current status. *ONLINE Conference*, London.

15. IEEE Standards, for Local Area Networks (1985). *Logical Link Control.* Wiley-Interscience, New York.

16. IEE Standards for Local Area Networks (1985). *Carrier Sense Multiple Access with Collision Detection (CSMA/CD) Access Method and Physical Layer Specificationss.* Wiley-Interscience, New York.

17. IEEE Standards for Local Area Networks (1985). *Token-passing Bus Access Method and Physical Layer Specification.* Wiley-Interscience, New York.

18. IEEE Standards for Local Area Networks (1985). *Token-passing Ring Access Method and Physical Layer Specifications.* Wiley-Interscience, New York.

10.5.9 Chapter 9

1. Hughes, C.J. (1984). Switching — state-of-the-art. *Br. Telecom Technol. J.* **4**, (1), 5–19 and **4**, (2), 5–17.

2. Houldsworth, J. (1984). Convergence of LAN and digital telephone exchange. In *Information Technology and the Computer Network*, Ed. Beauchamp, K.G. Springer-Verlag, Berlin.

3. CCITT Recommendation I.120 (1984). *Integrated Services Digital Network*, The Red Book, Vol. III.5. ITU, Geneva.

4. CEPT Special Group ISDN (GSI) (1984). Report on Integrated Services

Digital Network Studies, Carb 2, *Survey of Plans for the introduction of ISDN in Europe*, Document T/GSI (84) 37.

5. CCITT Recommendation I.412 (see ref. 3 above).
6. CCITT (1984). Network message handling systems. *The Red Book*, Vol. III.7. ITU, Geneva.
7. Fosdick, H. (1985). Initial experiences with multi-media documents. *Proc. 1st Int. Symp. Comput. Meas. Systems*. North-Holland, pp. 3–12.
8. Intercept Recommendations for Message Handling Systems (1985). *Techn. Guide TG102/1*, Dept of Trade and Industry, UK.
9. I E E Colloquium (1986). *Planning for Automated Manufacture*, Savoy Place, London WC2R 0BC, UK.

APPENDIX

SUMMARY OF MAJOR CCITT STANDARDS RECOMMENDATIONS

V-SERIES RECOMMENDATIONS FOR DATA TRANSMISSION OVER TELEPHONE CIRCUITS

Equivalent standards are shown in brackets.

V3	International alphabet numbering system
V4	General structure of signals in International Alphabet No. 5 for data transmission over public telephone networks
V5–V110	**Data transmission services and facilities**
V5	Modulation rates and date signalling rates for synchronous data transmission in a general switched network
V10 (RS-423)	Electrical characteristics for balanced double current interchange circuits
V11 (RS-422)	Electrical characteristics for unbalanced double current interchange circuits e.g. a twisted pair
V15	Use of acoustic couplers for data transmission
V21	Full duplex serial asynchronous transmission of digital data at speeds up to 300 bps for use in the public telephone network
V22	Full duplex serial transmission of digital data at 1200 bps for use in the public telephone network
V22bis	Full duplex serial transmission of digital data at 2400 bps for use in the public telephone network
V23	Half-duplex serial transmission of digital data at 600/

	1200 bps for use in the public telephone network
V24 (RS-232c)	Definitions for interchange circuits between a DTE and DCE (i.e. modem)
V25	Auto calling/answering equipment on the public telephone network
V26	Half-duplex serial transmission of digital data at 1200 or 2400 bps for use in the public telephone network (generally for leased point-to-point circuits)
V26bis	Half-duplex serial transmission of digital data at 1200 or 2400 bps for use in the public telephone network (generally for public lines)
V26ter	Full duplex synchronous or asynchronous transmission of digital data at 2400 or 1200 bps for use in the public telephone network
V27	Half-duplex serial transmission of digital data at 4800 bps for use in the public telephone network
V27bis	Half-duplex serial transmission of digital data at 4800/ 2400 bps for use in the public telephone network (generally for leased point-to-point circuits)
V27ter	Full duplex synchronous or asynchronous transmission of digital data at 4800 or 2400 bps for use in the public telephone network
V29	Full duplex synchronous transmission of digital data at 4800 or 9600 bps on double dial-up line or half-duplex on one telephone call in the public telephone network
V32	Full duplex synchronous or asynchronous transmission at 4800 or 9600 bps for use in the public telephone network
V33	Full duplex synchronous or asynchronous transmission at 14.4 kbps for use in the public telephone network
V35 (RS-449)	Interface definition between a DTE and DCE using electrical signals defined by V11
V36	Modems for synchronous data transmission using 60-- 108 kHz group band circuits
V52–V57	**Line testing**
V52	Burst error rate test for modems
V54	Loop test devices for modems

X SERIES RECOMMENDATIONS FOR DATA NETWORKS

X1–X15	**Services and facilities**
X1	International user classes of service in public data networks
X2	Packet-switched services (user facilities) in public data networks
X3, X28, X29	**Triple X facilities**

X3	Facilities for packet assembly/disassembly in public data networks
X4 (V4)	General structure of signals in International Alphabet No. 5 for data transmission over public data networks
X15	Services and facilities

X20–X32	**Interfaces**
X20	Interface between a DTE and DCE for start–stop (asynchronous) transmission on public data networks
X20bis	V21 compatible interface between DTE and DCE for start–stop transmission on public data networks
X21	Interface between DTE and DCE for synchronous operation on public data networks
X21bis	Used to access public data networks with analogue circuits, i.e. synchronous V-series modems
X24	Definition of interchange circuits between DTE and DCE on public data networks
X25	Interface between DTE and DCE for terminals operating in packet mode on public data networks
X26 (V10)	Electrical characteristics for unbalanced double current interchange circuits
X27 (V11)	Electrical characteristics for balanced double current interchange circuits, e.g. a twisted pair
X28	Interface between DTE and DCE for a start–stop mode data terminal accessing the PAD facility on a public data network
X29	Procedures for exchange of control information and user data between a packet mode DTE and a PAD
X32	Interface procedure for dial-up access to packet mode DTEs and teletext terminals from the PSTN

| **X40–87** | **Transmission, signalling and switching** |
| X75 | Packet mode protocol for inter-networking between different X25 networks |

X92–X141	**Network aspects**
X92	Hypothetical reference connections for synchronous public data networks
X95	Network parameters in public data networks
X96	Call progress signals in public data networks
X121	International numbering plan for public data networks
X150	Maintenance

| **X180–X181** | **Administrative arrangements** |

| **X200–X244** | **Open systems interconnection (OSI)** |

X200	Open systems interconnect — basic reference model
X210	OSI service definition conventions
X213	OSI network layer service
X214	OSI transport layer service
X215	OSI session layer service
X224	OSI transport layer protocol
X225	OSI session layer protocol
X250	System description techniques

X300–X310	**Interworking between networks**

X350–353	**Mobile data transmission systems**

X400–X430	**Message handling systems**
X400	Message handling system — service elements
X401	Message handling system — service elements and optional user facilities
X408	Encoded information type conversion rules
X409	Presentation transfer syntax and notation
X410	Remote operation and reliable message transfer service
X420	Electronic mail content protocol providing interconnection between different mail systems
X430	Access protocol for Teletext terminals

G-SERIES RECOMMENDATIONS FOR TRANSMISSION SYSTEMS AND MULTIPLEXING EQUIPMENT CHARACTERISTICS OF DIGITAL NETWORKS

G701–G941	**Interfacing of terminals with digital transmission networks**
G703	Physical, functional and electrical characteristics of primary PCM operation at 2.048 Mbps
G732	Principal characteristics of primary PCM operation at 2.048 Mbps

I-SERIES RECOMMENDATIONS FOR AN INTEGRATED SERVICES DIGITAL NETWORK

I110–I230	**General**
I110	Structure of the I-series recommendations
I111	Relationship with other recommendations relevant to ISDN
I112	Vocabulary of terms for ISDN

I210–212	**Service capabilities**
I210	Principles of telecommunication services supported by an ISDN

| I211 | Bearer services supported by an ISDN |
| I212 | Teleservices services supported by an ISDN |

I310–I335	**Overall network aspects**
I310	ISDN network functions and principles
I320	ISDN protocol reference and model
I330	ISDN numbering and addressing principles
I340	ISDN connection types

I410–I464	**User network interface**
I410	General aspects and principles relating to recommendations in ISDN user network interfaces
I411	ISDN user network interface — reference configuration
I412	ISDN user network interface — channel structure and access capabilities
I420	ISDN basic rate user access network interface
I421	ISDN primary rate user access network interface
I430	ISDN layer 1 specification for basic rate access user network interface
I431	ISDN layer 1 specification for primary rate access user network interface
I440	ISDN layer 2 user network interface — general
I441	ISDN layer 2 user network interface — specifications
I450	ISDN layer 3 signalling procedures for user network interface — general aspects
I451	ISDN layer 3 signalling procedures for user network interface — specifications
I461	Support for X21 and X21bis DTE by an ISDN
I462	Support for packet mode terminal equipment by an ISDN
I463	Support of V-series type interfaces by an ISDN
I472 (X31)	Packet modem terminal support in ISDN

ANSWERS TO PROBLEMS

CHAPTER 1

A1.1 **In-band signalling** is applied when both message information and signalling information are carried over the same line, e.g. in the public telephone service.

Out-band signalling (or common-channel signalling) is when message and signalling information are carried separately. A common channel thus conveys the signalling requirements of several message channels. This is applied in the ISDN public service.

A1.2 **Space division switching** applies to multi-wire transmission — one pair of wires per channel.

Frequency division multiplexed switching (FDM) and **time division multiplexed switching** (TDM) are applied to send several signals along a common transmission medium. FDM uses a fraction of the available frequency bandwidth for each signal whereas with TDM a fraction of the available transmission time is used.

A1.3 Let V_s = the signal level and V_n = the noise level = 20 mV. The **signal-to-noise ratio** (SNR) is given as 20 dB so that V_s = 200 mV. Line attenuation causes a reduction in signal level from an initial value of V_s' = 500 mV to the 200 mV level. This is equal to a power ratio of 8 dB giving a maximum length of line as 8 km.

A1.4

Power level	dBm
$1\mu W$	−30
$10\mu W$	−20
$100\mu W$	−10
1 mW	0

| +30 | 1 W |
| +40 | 10 W |

CHAPTER 2

A2.1 **a** From Eqn 2.2 and noting that the power of a signal is proportional to the square of its amplitude value averaged over the cycle interval time then:

$$P_T = \tfrac{1}{2}(A_c)^2 + \tfrac{1}{2}[\tfrac{1}{2}(m.A_c)]^2 + \tfrac{1}{2}[\tfrac{1}{2}(m.A_c)]^2$$

The carrier power is given as:

$$P_c = A_c^2/2$$

with the side-band power as:

$$P_s = (m^2.A_c^2)/4$$

and the ratio is:

$$P_s/P_t = m^2/(2 + m^2)$$

b For SSB modulation the total power is:

$$P_t = \tfrac{1}{2}(A_c)^2 + \tfrac{1}{2}[\tfrac{1}{2}(m.A_c)]^2$$

The side-band power is:
$$(m^2.A_c^2)/8$$

The ratio is:

$$m^2/(4 + m^2)$$

A2.2 Inspection of Fig. 2.6 shows that tolerance to error is determined by the proximity of one point in the constellation to its nearest neighbour through drawing a circle around a given point ensuring that this does not touch other circles similarly drawn. Both constellations are drawn on the same scale so that the comparative error is related to the diameter of the circles. From Fig. 2.6 an approximate ratio of 1.24 or 1.87 dB is obtained.

A2.3 This is shown in Fig. A1. The speech channel covers a range of 300–

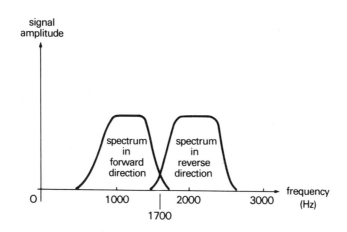

Fig. A1. Answer to problem P2.3.

3400 Hz and in full duplex operation signals are transmitted in both directions. Hence for each direction of transmission the bandwidth is divided into two regions each of 1700 Hz as shown. Refer to Table 2.2 for applications of these frequency bands.

A2.4 To avoid problems in completely filtering out the carrier signal, particularly with wideband signals, such as television.
To permit the carrier frequency to be derived from the transmitted signal for synchronization purposes.

A2.5 Your answer should include the improvement in transmission efficiency obtained through the use of a variable time slot for the data, and the technique of replacing empty data channels with a flag bit. Mention should also be made of the value of using statistical multiplexers in a concentrator design whereby the flag bits are removed for transmission, so compressing the data.

A2.6 The input to the demodulator part of a **modem** is an analogue signal modulated by a digital signal. The input to the coder part of a **codec** is an analogue signal to be represented as a sampled and quantized digital signal and modulation is not involved. Thus they do not do the same job.

CHAPTER 3

A3.1 a Advantages of using optical fibres are discussed in the text. A major practical difficulty lies in joining two cable to achieve minimum insertion loss.

b Referring to Fig. 3.2, since $\Theta_1 = \Theta_2$ and $b = p$ then $\cos \varnothing = b/h = \sin \Theta = p/h$

A3.2 **a** The pulse spreading delay is the bit separation for 2 Mbps, i.e. 0.5 μs. Since $n_1 = n_2$ then from Eqn 3.1: $t = L(n_1 - n_2)/c$. Substituting values you get $L = 7.5$ km.

b Loss = dB times length = 30 dB.

A3.3 **a** For a graded fibre from Eqn 3.2: $t = L/2c.(n_{max}\ d^2)$, with $d = 0.01$, $n_{max} = 1.5$, $L = 10\ 000$ m, then $t = 0.25$ μs. For a multimode fibre then from Eqn 3.1, $t = 1.67$ μs.

b Pulse spreading with a graded fibre compared with a multimode fibre of similar characteristics is smaller and this enables shorter pulses to be transmitted without distortion and hence increase the maximum rate of transmission.

A3.4 Bi-directional communication is achieved by using different frequencies for upwards transmission than for downwards transmission.

A major difference between speech transmission by satellite and by the telephone network is the delay in satellite transmission of approximately 250 ms. This causes difficulties in speech dialogue.

A3.5 Signal-to-noise ratio (SNR) is defined in the text. Total Gaussian noise is: $n = 10 + 100 + 10 = 120$ mW, and at peak times $n = 150$ mW. Therefore the SNR for a 500 mW signal is:

a Generally $\qquad 10 \log_{10} 500/120 = 6.2$ dB.

b At peak times; $10 \log_{10} 500/150 = 4.8$ dB.

A3.6 From Eqn 2.7 the channel capacity in bps is:
$C = 50\ 000 = B \log_2 (1 + s)$.
The phase jitter noise is 10 dB and expressing this as a ratio, $S = 10$, so that $B = 14\ 450$ Hz.

A3.7 With any transmission system the signal is attenuated as it propagates along the medium. Amplifiers/repeaters are used to restore the signal back to its transmitted level. An amplifier is used to restore analogue signals. A repeater does the same for a digital signal. Without the use of these the signal would attenuate until it is so small compared with the noise present on the line as to render the transmission useless. A limitation in the number of amplifiers/repeaters is set by the transmission impairments, principally amplifier non-linearity in the case of analogue signals and cumulative noise for digital repeaters.

CHAPTER 4

A4.1 The message, k=1101011. The divisor, p=11001.
a The augmented message has four zeros added to give: k'=11010110000. Dividing this augmented message to derive a remainder, using binary division gives: r = 1010. Hence the transmitted frame is 11010111010.
b Decoding this through binary division by r gives a remainder of 000000 indicating no errors.
c Adding an error burst of three digits results in a transmitted message of 11010111101. Dividing this by p gives a remainder of r = 000111 indicating an error in the last 3 digital places.

A4.2 **a** A major advantage is improved resolution obtained for the lower frequencies in the transmitted speech. Companding advantage is expressed as a logarithmic ratio of the number of sampling levels possible with uniform and non-uniform quantizing for the same dynamic range
b A plot of x against y similar to that shown in Fig. 4.3 is required for 1024 decision values. The number of non-uniform decision values is given as 128 (7 bits).
c The companding advantage is therefore 18.0 dB.

A4.3 The main advantage of bit-oriented protocols are:
because control information is at the bit rather than the character level, there are fewer overhead bits per frame; and
for the same reason, it is easier to add new features.

See test for further discussion

A4.4 The sampling value represents the average value of the signal over the sampling period. Too few such periods and the fine details of the signal are lost. Two ways of minimizing this are the use of very fast digitizing techniques and the use of a sample-and-hold device to maintain the amplitude of the sample constant during the quantizing process.

A4.5 **a** The sampling rate f_s is derived from the highest angular frequency present giving f_s = 318 Hz.
b The combined peak value of the three consinusoidal waveforms is $\pm(3 + 10 + 14) = \pm27$ giving a peak-to-peak signal of 54. With N=10 bits there are 1024 quantization levels, so that the value of the quantizing interval is: $54/1024 = 0.053$.

Substituting in eqn 4.2 gives the SNR as: SNR = $6 \times \log_2 1024 + 1.8 = 61.8$ dB.

A4.6 From eqn 2.7 the channel capacity is theoretically $C = 3000 \log_2 (1 + R)$, where $S = 20 = 10 \log_{10} R$ so that $C = 19$ dB. However, a band-limited channel of 3 kHz bandwidth will only support a sampled signal of up to 6 kbps giving a limitation of 6 kbps on the attainable transmission speed.

A4.7 Null modem interconnections are shown in Fig. 4.9. These interconnections can be related to the RS-232c interface signals shown in Table 4.1. Thus the transmitted data connection, BA, at one end of the null modem becomes the received data connection, BB, at the other end and so on.

CHAPTER 5

A5.1 Total set-up time $= (50 \times 20)/1000 = 1$ s. Propagation delay for request and acknowledgement signals is the round trip divided by the propagation velocity $= 5$ ms.
Time to transmit the message is:

    a 100 s,
    b 100 ms.

Therefore the total time to transmit the message is:

    a $1 + 0.005 + 100 = 101$ s,
    b $1 + 0.005 + 0.1 = 1.1$ s.

The efficiency of the circuit switching network is poor for short messages where the time to set up the network forms a major part of the total time.

A5.2

Let efficiency $A = \dfrac{\text{Propagation time } (T\text{p})}{\text{Transmission time } (T\text{m})}$

The propagation delay is given as 20 ms.

Transmission time. $T\text{m} = \dfrac{\text{length of the message, } M}{\text{data rate (bps)}}$

$$= M/400$$

so that $\qquad A = \dfrac{20 \times 400}{1000 \times M} = \dfrac{1}{2}$ (50%).

Therefore the minimum message length = 160 bits.

A5.3 Transmission delay = 52 s. Total time required is 52 + 5 + 50 = 107 s. Therefore the effective data rate is 50 000 bits in 107 s, i.e. 4672 bps.

A5.4 Applying the optimization procedure outlined in section 5.8:
    a Given the channel capacity of $C = 9600$ bps and average packet size of 500 bits = $1/a$ then a $C = 19.2$. Substituting this in eqn 5.3, the delay for a given link

$$T_i = \frac{1}{19.2 - X_i} \text{ seconds}$$

X_i is different for each link in the network chain. Given a mean packet arrival rate of $A = 50$ per second then from eqn 5.4 the overall average delay time is:

$$T = \sum_{i=1}^{3} \frac{X_i}{50} \; \frac{1}{19.2 - X_i}$$

Substituting values given for X_i then: $T = 0.02 + 0.07 + 0.03 = 0.12$ s.

b The minimum network capacity is based on the highest number of packets on the link and their average size. Thus for link B–C this is 15 packets per second and the packets have an average length of 500 bits. The minimum network capacity is $15 \times 500 = 7500$ bps which is less than the stated capacity of 9600 bps, showing that the network is viable.

A5.5 **Time-out** is a general term used for a timing operation. At the commencement of a process a timer is set into operation. At a given number of cycles set by the time-out parameter, the time-out is reached and a specific action precipitated (e.g. a data packet may be transmitted).
    Isarithmic flow control and **stop-and-wait** flow control are described in section 5.7.

A5.6 A time of 500 ms is available to transmit the 960 bits. Hence a minimum rate of 1920 bps is needed. (Slightly larger would be allowed for differences between access and transmission delay.) Storage of $2 \times 960 = 1920$ bits is required to permit repetition in the event of transmission failure or corruption.

CHAPTER 6

A6.1 **a** A wide area network generally takes the form of a mesh topology containing a series of interconnected nodes to give a number of alternate routes across the network. A local area network encompasses a smaller geographical area and will take the form of a bus or ring topology. Here data are transferred along a restricted bath containing all the nodes of the network with no alternative routing except (in some cases) that of direction of data flow.
b The speed of operation for a wide area network is considerably lower than that found in a local area network. The operational problems for the former are concerned with routing, traffic control and congestion, whereas in the local area network it is necessary to consider contention between users sharing the same media, and prevention of continued repetition of the message throughout the network once it has been correctly received.

A6.2 In a bus network the nodes behave as transceivers and act in a passive role receiving all messages entering the network. A transceiver is only activated to pass on the data to its linked end device when it recognizes the device address contained in the data. In a ring network the node also acts as a repeater, transmitting to the next node data presented at its input.

A6.3 A value added network is a carrier communication network to which is added computer control of the data such that a number of user services can be provided as well as transmission of the message data.
The advantages of such an enhanced network to the user are:

1 only a small capital investment is required to gain access to the network;
2 the organization providing the value added network takes full responsibility for its development and maintenance; and
3 tariff charges are for actual connection time the network.

A6.4 **a** With no data to send, the poll propagates at a constant rate from station to station through the network. Each inter-station transmission takes 6.67 ms. To this is added the modem time of 10 ms to give 16.67 ms or 60 polls per second.

b Given eight secondary stations, each link requires 16.67 ms, i.e. 133.4 ms total. The time to send a 100-character message back to the primary station is 800/9600 = 83 ms. Thus the total time is 133 + 83 + 216.4 ms.

A6.5 **a** A terminal-switched exchange provides connection between a terminal and a computer handling digital data. A CBX, in addition, can handle digitized speech connection including control signalling for telephone connection, i.e. ringing, speech interchange and standdown.
b A speech signal has a bandwidth of approximately 4 kHz. Sampling at 8 kHz using an 8-bit character to describe each sample, a minimum data rate of 8 × 8 = 64 kbps is required.

A6.6 This depends on the ratio of the two transmitted speeds, the storage available at each node, and the spacing between messages. With a lengthy period between messages the limitation depends on the ratio of input/output transmission speed and the minimum storage availability at the nodes.

CHAPTER 7

A7.1 **a** Bus: Since the bus network is a passive broadcast device, a single node failure should have no effect the network operation. However, arrangements are usually made to disconnect the node in the event of failure.
b Ring: Each node in a ring configuration is responsible for passing on (repeating) information present on the ring to the next node in the transmission path of the network. Consequently, a failure of one node could seriously interrupt the flow of data through the network.
c Star: The failure of a peripheral node would have no effect on the network for other peripheral nodes. Failure of the hub node would, however, render the entire network unusable since all data passes through this node.

A7.2 The bit length of a link between two stations in the network is:

$$\text{bit length} = \frac{\text{data rate} \times \text{inter-section spacing}}{\text{propagation velocity}}$$

Substituting values: for the slotted ring this is 1.25 bits; for the token ring this is 0.725 bits. Assuming a delay at each repeater station to be 1 bit, then the total bit length is:

2.25 × 200 = 450 bits for the slotted ring
1.725 × 200 = 345 bits for the token ring.

Hence, the number of slots that can be accommodated in the network is 450/50 = 9, and the number of tokens is 345/8 = 43.

A7.3 Transmission time = 512/10 = 51.2 μs.
Acknowledgement time = 3.2 μs.
During this total time of 54.4 μs a total of 512 – 32 = 480 bits of data are sent.
Therefore the data rate is = 8.82 Mbps.

A7.4 Propagation time around ring is:

$(350 × 10^{-7}) + (10^4 /200 × 10^6) = 85$ μs.
Therefore the bit length of the ring is = 850 bits, and the number of slots is: 850/40 = 21 shots.

A7.5 **a** Definitions for a baseband LAN and a broadband LAN are given in the text.
b Duplex operation in a broadband system is obtained by allocating different frequencies for 'go' and 'return' paths. Bi-directional transmission is not possible since unidirectional analogue amplifiers are used at the nodes to pass on the signal.

A7.6 Star-ring architecture enables new stations to be added and faulty stations to be removed from the network at a central point. A major disadvantage is the additional cable required to connect each node to the central point.

CHAPTER 8

A8.1 **a** The seven-layers and their functions are described in the text.
b With LAN communication it is necessary to transmit a stream of digital bits accurately across the network. To do this not only must the physical specification of the signals constituting the bits be defined (ISO layer 1) but also the beginning and end of each block of bits needs defining and means included to correct errors (layer 2). Some attributes of layer 3 are required such as the need to handle multiple data streams. It is usual to redefine the two lowest layers plus part of layer 3 into two LAN-defined layers. These are termed the medium access control layer (MAC) and a logical link control layer (LLC), both of which are described in the text.

A8.2 This process was described in section 8.4.1 using Fig. 8.8 as an explanatory diagram.

A8.3 **a** Normal and asynchronous response modes are described in the text. They are defined through the HDLC frame by the contents of the control field.
b Flow control is obtained through the information frames of the control field. This applies the sliding window technique by utilizing two parameters, SEQ and NEXT, which tell the receiver the sequence numbers of packets received and the packet number expected in the next transmission.

A8.4 The function of the LLC and MAC layers are defined in section 8.4. Their relationship to the OSI layers is given in Fig. 8.9.

A8.5 Let T_1 = time to transmit a frame = 1.024 ms.
The round-trip time to include acknowledgement is: 270 + 1.024 + 270 = 541 ms. In an HDLC frame, 48 bits are reserved for control and overheads so that the data actually sent are: 1024 – 48 = 976 bits. Therefore, maximum transmission rate is 976/541 = 1.8 kbps.

A8.6 A general description of sliding window flow control is given in section 5.7.2 with reference to Fig. 5.7. The SEQ and NEXT bits in the control field of the HDLC frame (Fig. 8.5) would be used to indicate the order of the packets transmitted and to indicate the next frame number to be expected (the SEQ bits), and when an acknowledgement to a series of packets should be sent (the NEXT bits). Taken together, this mechanism will control the reassembly of packets in their correct message order.

CHAPTER 9

A9.1 **a** With alternate mark inversion coding, the polarity of alternate pulses representing 1s is automatically changed before transmission. If this fails to occur (by deliberate violation) then either a pulse will be omitted or falsely added. Either of these events will result in two successive pulses being of the same polarity — an event which can easily be detected by circuit logic.
b The contention bit can be changed from, say, a 0 bit to a 1 bit by the station transmitting a message. Terminals will contend for use of the channel by reading this bit position and if a 1 is read will defer transmission until a later period. This situation is very similar to the CSMA/CD procedure described in Chapter 7.

A9.2 The rationale for a seven-layer protocol in MAP communications protocols is outlined in section 9.3. Token bus is preferred to contention protocol, such as CSMA/CD, on the grounds of high reliability — an essential feature in the manufacturing process.

A9.3 The operation of a message transfer agent and user agent in an electronic message system is given in section 9.2.1. A resumé of these procedures needs to be given in an answer to this question.

A9.4 The bandwidth of a primary rate access channel is 2.048 Mbps and consists of 30 channels of 64 kbps data (the B channel) and a control channel of 64 kbps of low-priority data and control (the D channel). The traffic to be sent is: 80 telephone channels, each of 64 kbps.

Three slow TV channels where each frame is to be completed in 2 s. In this time approximately $600 \times 600 \times 4/5 = 288$ kbps are to be sent on each channel. This equals 432 kbps, equivalent to: 7 channels, each of 64 kbps.

50 data channels of 2.4 kbps requiring: 2 channels, each of 64 kbps.

150 data channels of 1.2 kbps and of low priority requiring: 3 channels, each of 64 kbps.

This gives 89 high-priority channels and 3 low-priority channels so that three primary rate access channels would be needed with the B channels allocated to the high-priority signals and the D channels allocated to the low-priority signals.

A9.5 The LAN for a MAP must be highly reliable under all loading conditions since on-line manufacture processes are being controlled. This rules out contention protocol techniques which can include an arbitrary delay. A deterministic system, such as a token ring/bus system is therefore chosen. A TOP system can accommodate small and variable delays since non-critical office procedures are involved. Hence this can use the cheaper contention protocols such as CSMA/CD.

INDEX